HYPERFINE INTERACTIONS (C)

PROCEEDINGS OF LACAME '98

Sixth Latin American Conference on Applications of the Mössbauer Effect

Cartagena de Indias, Columbia
September 13 – 19, 1998

EDITORS:
Jesús A. Tabares, Chairman
Alberto Bohórquez G, Coordinator
Ligia E. Zamora
Germán A Pérez Alcázar

BALTZER SCIENCE PUBLISHERS, BUSSUM, THE NETHERLANDS

PRINTED IN THE NETHERLANDS BY THIEME DEVENTER BV

Organizing Committee

Germán A. Pérez Alcázar - *Chairperson*
Ligia E. Zamora - *Coordinator*
Jesús A. Tabares
César Augusto Barrero
Alvaro Morales
Carlos Arroyave

Latin American Executive Committee

Roberto Carlos Mercader (ARGENTINA)
Celia Saragovi Badler (ARGENTINA)
Elisa Baggio-Saitovich (BRAZIL)
Germán A. Pérez Alcázar (COLOMBIA)
Edilso Reguera Ruiz (CUBA)
Hemani Yee-Madeira (MEXICO)
Víctor A. Peña Rodríguez (PERU)
Fernando González Jimenez (VENEZUELA)

Program Committee

Bibiana Arcondo (ARGENTINA)
Wagner A. Macedo (BRAZIL)
Rosa B. Scorzelli (ARGENTINA)
César A. Barrero (COLOMBIA)
Jesús A. Tabares Giraldo (COLOMBIA)
Juán A. Jaén (PANAMA)
Fernando González Jiménez (VENEZUELA)

Local Committee

Ligia E. Zamora
Oriana Benavides
Laura A. Herrera
Germán A. Pérez Alcázar
Alberto Bohórquez G.
Jesús A. Tabares

Proceeding Committee

Jesús A. Tabares - *Chairman*
Alberto Bohórquez G - *Coordinator*
Ligia E. Zamora
Germán A. Pérez Alcázar

Acknowledgments

This Conference has been sponsored and organized by the Physical Departments of the Universidad del Valle in Cali and of the Universidad de Antioquia in Medellín.

Financial support has been provided by the following organizations:

CLAF - Centro Latinoamericano de Física
ICTP - International Centre for Theoretical Physics
SECAB - Secretaría Ejecutiva del Convenio Andrés Bello
ICETEX - Instituto Colombiano de Educación Técnica en el Exterior
ICFES - Instituto Colombiano para el Fomento de la Educación Superior
COLCIENCIAS - Instituto Colombiano para el Desarrollo de la Ciencia y la Tecnología, Francisco José de Caldas
SOCOLFI - Sociedad Colombiana de Física
UNIVALLE - Universidad del Valle (Fundación Multitaller, Fundación de Apoyo, Research Vice-rectoría and Physics Postgraduate Program)
UdeA - Universidad de Antioquia

This publication was supported from COLCIENCIAS, whose main aid is to improve the Colombian scientific and technological development.

PREFACE

The Sixth Latin American Conference on Applications of the Mössbauer Effect (LACAME'98) was held at the Caribe Hotel in Cartagena de Indias, Colombia, during September 13-19, 1998. The program consisted of 10 invited talks, 28 oral presentations and 60 contributed paper in the form of posters. 61 registered participants from 13 countries attended the meeting, 50 of them from Latinoamerica including 23 undergraduate and graduate students.

All the presented contributions were related to the principal applications of the Mössbauer Effect but for the invited talks four topics were selected: Soils and Mineralogy, Amorphous Materials and Nanocrystalline Materials, Magnetism and Magnetic Materials and Corrosion and Surface Phenomena. These topics are those in which the majority of the Latin American community are working.

The proceedings of the conference are published in two volumes: the one with the invited lectures and oral presentations is a normal issue of this journal and the other, with the other contributions, is a special issue C for conferences.

After the Latin American scientists Meeting, held at the end of the conference, the community agreed that these conferences are very important in order to increase the interactions and infrastructure and are committed to organize the seventh Conference LACAME"2000 in Caracas, Venezuela.

April, 1999   Jesús Anselmo Tabares Giraldo (editor, chairman
                     Alberto Bohórquez Gallo (coordinator)

Contents

MÖSSBAUER STUDIES ON HETEROBIMETALLIC COMPLEXES BEARING AT LEAST ONE RUTHENIUM-TIN σ-BOND

Edmilson M. Moura[1], Michael H. Dickman[2], Gennaro J. Gama[1,*], Helmuth G.L. Siebald[1,5 *], Anuar Abras[3] and Manfredo Hörner[4]

[1]Departamento de Química, ICEx, Universidade Federal de Minas Gerais, Belo Horizonte, MG - 31270-901 Brazil
[2]Department of Chemistry. Georgetown University, Washington, DC 20057-2222 USA
[3]Departamento de Física, ICEx, Universidade Federal de Minas Gerais, Belo Horizonte, MG - 31270-901 Brazil
[4] Departamento de Química, Universidade Federal de Santa Maria, Santa Maria, RS, 97119-900 Brazil

A series of heterobimetallic complexes containing at least one Ru-Sn σ-bond, with general formula $[RuCp(L)_2SnX_2Y]$ (L = PPh$_3$, triphenylphosphine; ½ ddpe =1,2-bis(diphenylphosphino)ethane; X,Y = F, Cl, Br), was prepared and studied by elemental analysis, X-ray diffraction, Mössbauer spectroscopy, I.r. and (^{1}H, ^{13}C, ^{19}F, ^{31}P and ^{119}Sn) NMR spectroscopy. Mössbauer studies allowed the determination of the number of coordination around Sn atom as well as the electronegativity of the organometallic fragment $[RuCp(L)_2]^+$. These results, with the support of the NMR data, allowed the inference of the effect of the nature of L, X and Y on the electronic distribution between the Ru and Sn centers and the effect of L on the electronegativity of the organometallic fragment.

1.Introduction

Organoheterobimetallic compounds containing a transition-metal to tin bond are of special importance in the field of catalysis and as models for the study of dinitrogen fixation[1]. Of special interest are those air-stable heterobimetallic complexes containing the cyclopentadienyl diphosphine ruthenium(II) fragment, CpRu(PR$_3$)$_2$$^+$, which are frequently used in several synthetic procedures. For instance, the facile replacement of both the phosphines or of the chloride in (η^5-C$_5$H$_5$)Ru(PPh$_3$)$_2$Cl by a wide variety of neutral or anionic ligands has been used [2] as a means to prepare several complexes containing a metal-bound organic moiety. These compounds have been used to promote modifications of such organic fragments even under mild conditions [3-5]. Interest in phosphino complexes of Ru(II) also stems from their catalytic activity on processes such as the decarbonylation of both aliphatic and aromatic aldehydes as well as in hydroformylation reactions carried out under mild conditions [6-9].

The catalytic activity of $[(\eta^5$-C$_5$H$_5$)Ru(PPh$_3$)$_2$X]$, X= Cl, SnCl$_3$ and SnF$_3$, on the low-temperature conversion of methanol into methyl acetate poses an interesting dichotomy as those compounds with X= Cl, SnCl$_3$ show very low activity while for the complex with X = SnF$_3$, the number of catalytic turnovers is *ca.* 80 times higher than those displayed by the former complexes [9]. It then becomes clear that not only Sn must be present in the composition of the catalyst but also that subtle changes of the coordination environment around the tin center, and therefore the electronic distribution along the Ru-Sn bond, must exert a strong influence on the catalytic activity of these complexes. The investigation of the dependence of the electronic distribution along the structure of compounds structurally related to $[(\eta^5$-C$_5$H$_5$)Ru(PPh$_3$)$_2$X]$ on the nature of the ligands around both metal centers is of

---

* Corresponding authors.[5]Permanent address: Departamento de Farmácia. Universidade de Uberaba (UniUbe) Uberaba, MG - 38065-500 Brazil

fundamental importance to the understanding of their potential catalytic activity since substrate activation processes are extremely dependent upon this distribution.

The electronic distribution along the Ru-Sn bond will be the product of the competition for the Ru-based and Sn-based electrons by the ligands around each metal atom. Electronic density will be shifted towards one specific center as the electronegativity, χ, of the moieties bound to it overcome that of the ligands around the other atom. Although it is simple to define the electronegativity of monoatomic ligands (e.g. F, Cl, Br), the inference of the same value for polyatomic ligands, such as organometallic fragments, is troublesome.

[119]Sn Mössbauer Spectroscopy proved useful in providing information regarding the chemical environment around tin atoms. From its use it is possible to infer both the electronegativity and the spatial arrangement of the moieties coordinated to Sn centers. This work reports Mössbauer studies carried out on compounds with the general formula $[RuCp(L)_2SnX_2Y]$ (L = PPh_3, ½ dppe; X = Y = F, Cl, Br; and dppe = $Ph_2P(CH_2)_2PPh_2$) or $[RuCp(dppe)SnX_2Y]$. It also describes the influence of the nature of L and of the manner at which dppe coordinates to Ru on the charge distribution between Ru and Sn is discussed.

2. Experimental

All reactions were carried out under standard Schlenck or dry-bag techniques under a dry-dinitrogen atmosphere. All solid products are air-stable, but solutions of compounds containing the SnF_3 moiety bound to Ru slowly hydrolyze in the presence of moisture to yield SnO_2 in addition to other products.

$RuCl_3 \cdot 3H_2O$ (Johnson Matthey Ltd.), anhydrous SnX_2 (X = F, Cl, Br. Aldrich Chemical Co.), NH_4X (X = F, Br. Aldrich Chemical Co.), 1,2-bis(diphenylphosphino)ethane (dppe. Aldrich Chemical Co.) were used as supplied. Cyclopentadiene dimer (Aldrich Chemical Co.) was distilled under N_2 prior to use. Ethanol was distilled from a $EtOH:Mg/I_2$ slurr under a flow of dry-N_2 and stored over 3Å molecular sieves. CH_2Cl_2 and $CHCl_3$ were distilled from a CaH_2 slurr and stored over 4Å molecular sieves. Benzene and toluene were distilled from a benzophenone ketyl still under dry-N_2 and stored over 3Å molecular sieves.

2.1. Syntheses of the Complexes

$(\eta^5\text{-}Cp)Ru(PPh_3)_2Cl$ and $(\eta^5\text{-}Cp)Ru(dppe)Cl$, the precursor compounds for the remaining syntheses, were prepared and purified according to published methods [10,11]. $[(CpRuCl)_2(\mu\text{-}dppe)_2]$ was separated from the reaction medium [12] for the synthesis of CpRu(dppe)Cl and used to prepare (5).

$(\eta^5\text{-}Cp)Ru(PPh_3)_2X$ compounds (X = F, Br) were prepared according to a slight modification of literature procedures [11,13] from the reaction between $(\eta^5\text{-}Cp)Ru(PPh_3)_2Cl$ and an excess of NH_4X:KX in refluxing EtOH:benzene (4:1).

$CpRu(PPh_3)_2(SnX_2Y)$, and $CpRu(dppe)(SnX_2Y)$, complexes were prepared from the reaction between $CpRu(PPh_3)_2Y$ or $CpRu(dppe)Y$ and SnX_2. A typical procedure is as follows:

$(\eta^5\text{-}Cp)Ru(PPh_3)_2Cl$ (0.371g - 5.1×10^{-4} mole) reacted with $SnCl_2$ (0.100g - 5.1×10^{-4} mole) in refluxing EtOH (100 mL) under a flow of dry-N_2. After 5 hours of reflux, the volume of the solution was reduced to ca. 30 mL under vacuum causing the precipitation of the product. The bulk product was separated by filtration and purified by dissolving in CH_2Cl_2 (4.0 mL) and inducing its precipitation by the addition of 20 mL of chilled n-hexane. After filtration, the solid was dried in vacuo at 35°C for at least one hour.

X-ray quality crystals of $CpRu(PPh_3)_2(SnCl_3)$ and $CpRu(dppe)SnBr_3$ were grown from the slow-evaporation of a $CHCl_3$ solution to which a small amount (ca. 10% volume) of n-hexane was added. Crystals were formed in 5 to 10 days.

2.2 Physical Measurements

Infra-red spectra were recorded with samples pressed as CsI pellets on a Perkin-Elmer 283B spectrometer in the 4,000 cm^{-1} to 200 cm^{-1} range. Carbon and hydrogen analyses were performed on a Perkin-Elmer PE-2400 CHN-analyzer using tin sample-tubes. Bromine and chlorine analyses were performed by means of X-ray fluorescence in a Rigaku-Geigerflex spectrophotometer with samples pressed as borax plates. Tin analysis was carried out using a Hitachi Z-8200 Atomic Absorption spectrophotometer (Sn lamp, λ_{max} = 224.6 nm. N$_2$O/acetylene flame).

^{119}Sn-Mössbauer spectra were collected in a standard instrument with samples kept at 77K using a CaSnO$_3$ source kept at room temperature. All isomer shift values reported in this work are given with respect to this source. All spectra were fitted by means of Lorentzian line shapes.

^{1}H, ^{13}C{^{1}H}, ^{31}P{^{1}H}, ^{119}Sn{^{1}H} NMR spectra were collected on a Bruker DRX400 spectrometer at 400, 50, 162 and 149 MHz, respectively and referenced to internal TMS (δ=0, ^{1}H and ^{13}C{^{1}H}), external 85% H$_3$PO$_4$/D$_2$O (δ = 0, ^{31}P{^{1}H}) and internal Sn(CH$_3$)$_4$ (δ=0, ^{119}Sn{^{1}H}).

Single crystal X-ray diffraction experiments were carried out on a Siemens P4 Smart CCD or an automatic four-circle diffractometer (Enraf-Nonius CAD4) using graphite-monochromated Mo-K$_{\alpha 1}$ (λ = 0.71073Å) radiation and an area detector. All the structures were solved and refined using the SHELX software[15].

3. Results and Discussion

Characterization data are summarized in Table 1. Elemental analyses allowed the determination of the empirical formulae as shown in that table. All I.r. spectra display the absorption characteristic of Sn-X stretching (X = F, Cl, Br), which are slightly shifted with respect to those observed for pure SnX$_2$. The disappearance of the absorption due to ν_{Ru-Cl} at 280 cm^{-1}, which was replaced by absorptions due to ν_{Sn-Cl}, ν_{Sn-Br}, ν_{Sn-F} at *ca.* 270cm^{-1}, 260 cm^{-1} and 490 cm^{-1} respectively, was observed for all pertinent cases, upon reaction of the precursors with SnX$_2$.

NMR results are summarized in Table 2. The chemical shifts observed, as well as the overall spectral patterns for all the experiments, together with peak-integral ratios in ^{1}H NMR spectra, point to the presence of discrete molecules in solution, at room temperature. If any self-assembly, such as oligomerization through shared Sn-X bonds between two or more molecules, were to take place in solution more complex NMR line-patterns as well as deviation from the ideal peak-integral values would have been observed. Exception to this general observation is observed in the ^{119}Sn NMR spectra of the SnF$_3$ derivatives, which show lines considerably broader than those for the other compounds.

The absence of self-association in the solid state for SnBr$_3$ and SnCl$_3$ derivatives was confirmed by single-crystal X-ray diffraction experiments, although the instability of SnF$_3$ derivatives, (6) and (7) , precluded any structural analysis for them. Therefore there is a considerable uncertainty regarding their oligomerization in the solid state at low temperatures but the considerably broad shape of I.r. bands due to Sn-F may be indicative of this self association. Compounds (1) and (4) are monomeric in the solid state at 173 K and X-ray data for (2) collected at 293 K also show this to be monomeric. The structures of (1) and (4) are shown in Figures 1 and 2, respectively.

Table 1 - Empirical Formulae, Color, Main I.r. Absorptions and Analytical Data for the Complexes

Complex, color	Empirical Formula	m.p (°C)	Main I.r. absorptions, cm⁻¹ (assignment)	Elemental Analysis (%) Found (calc.)
(1) bright yellow	$CpRu(PPh_3)_2SnCl_3$	188	290 (ν_{Sn-Cl})	C:54.31(53.77); H:9.95(9.85); Cl:11.20(11.61) Sn:13.23(12.97)
(2) red-orange	$CpRu(PPh_3)_2SnBr_3$	181	265, 260 (ν_{Sn-Br})	C:46.90(46.94);H:2.37(2.36); Br:22.71(22.85);Sn:11.56(11.31)
(3) bright yellow	$CpRu(dppe)SnCl_3$	255	291(ν_{Sn-Cl})	C:46.39(47.15);H:3.70(3.70); Sn:15.15(15.03)
(4) red-orange	$CpRu(dppe)SnBr_3$	257	256, 263 (ν_{Sn-Br})	C:41.10(40.30);H:2.87(3.17); Br:25.02(26.00);Sn:12.85(12.86)
(5) light orange	$\{[CpRu(SnCl_3)_2]_2(\mu\text{-}dppe)_2\}$	274	288(ν_{Sn-Cl})	C:47.14(47.15); H:3.34(3.70); Sn:15.07(15.03)
(6) pale-yellow	$CpRu(dppe)SnF_3$	271	493 (ν_{Sn-F}), *broad*	C: 51.15(50.30); H: 3.63 (3.95); Sn:15.05(16.03)
(7) light yellow	$CpRu(PPh_3)_2SnF_3$	199	493 (ν_{Sn-F}), *broad*	C:54.22(56.86);H:3.44(4.07); Sn:13.80(13.70)
(8) yellow	$CpRu(PPh_3)_2SnF_2Cl$	194-197	279 (ν_{Sn-Cl}), 498 (ν_{Sn-F})	C:54.75(54.55); H: 3.92(3.90); Sn: 13.20(13.57)
(9) light orange	$CpRu(PPh_3)_2SnF_2Br$	177-179	263 (ν_{Sn-Br}), 487 (ν_{Sn-F})	Sn: 12.14 (12.25); Br: 8.30 (8.22)
(10) orange-yellow	$CpRu(PPh_3)_2SnCl_2Br$	184-187	264 (ν_{Sn-Br}), 273 (ν_{Sn-Cl})	Sn:12.49(12.33);Br:15.72 (15.92)
(11) orange	$CpRu(PPh_3)_2SnBr_2Cl$	168-170	265 (ν_{Sn-Br}), 273 (ν_{Sn-Cl})	C. 49.01 (50.21); H: 3.51(3.31); Cl:3.53(3.80); Br:15.72(15.92)
(12) yellow	$CpRu(dppe)SnF_2Cl$	271-275	290 (ν_{Sn-Cl}), 500 (ν_{Sn-F})	C: 49.20(48.36); H:3.86 (3.52); Sn: 15.60(15.72)
(13) light orange	$CpRu(dppe)SnCl_2Br$	250-252	254 (ν_{Sn-Br}), 273 (ν_{Sn-Cl})	Cl: 8.51(8.20); Br: 9.59 (9.41); Sn: 14.25(14.07)
(14) orange	$CpRu(dppe)SnBr_2Cl$	251-253	252 (ν_{Sn-Br}), 270 (ν_{Sn-Cl})	C: 42.38 (40.50);H: 3.33 (2.59); Cl:4.04 (4.00);Br:18.19(17.80)

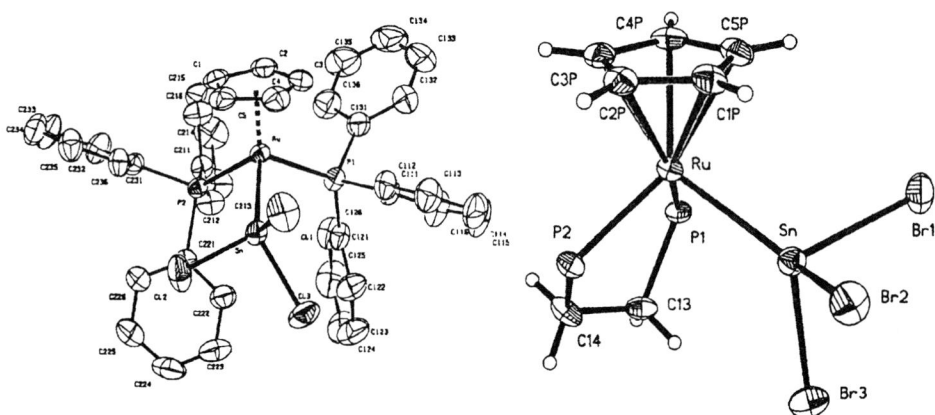

Figure 1 - Structure of $CpRu(PPh_3)_2SnCl_3$, (1).
Phenyl groups were omitted for clarity.
Relevant inter-atomic distances are: Sn-
Cl(1) = 2.4010Å; Sn-Cl(2) = 2.4017Å;
Sn-Cl(3) = 2.3800Å; Sn-Ru = 2.5720Å

Figure 2 - Structure of $CpRu(dppe)SnBr_3$, (4).
Relevant inter-atomic distances are:
Sn-Br(1) = 2.5478Å; Sn-Br(2) =
2.5659Å; Sn-Br(3) = 2.5407Å;
Sn-Ru = 2.5649Å.

Table 2 - ^{1}H, ^{13}C{^{1}H}, ^{31}P{^{1}H} and ^{119}Sn{^{1}H} NMR Results for the Heterobimetallic Complexes

Complex	Line Position, ppm. (multiplicity, assignment)			
	^{1}H	^{13}C{^{1}H}	^{31}P{^{1}H}	^{119}Sn{^{1}H}
(1)	4.52(s, C$_5$H$_5$); 7.01-7.37(m,C$_6$H$_5$)	81.4(s, C$_5$H$_5$); 127.9 –138.6(s,C$_6$H$_5$)	45.4(s, PPh$_3$)	-67.7(t, SnCl$_3$);
(2)	4.53(s, C$_5$H$_5$); 7.07-7.47(m, C$_6$H$_5$)	83.2(s, C$_5$H$_5$); 127.4-136.5(s,C$_6$H$_5$)	45.8 (s, PPh$_3$)	-139.2 (t, SnBr$_3$);
(3)	4.88(s, C$_5$H$_5$); 2.31-3.01(m, CH$_2$); 7.02-8.01 (m, C$_6$H$_5$)	80.4(s, C$_5$H$_5$); 27.8(s, CH$_2$); 128.3-133.1·(s, C$_6$H$_5$)	77.8 (s,dppe)	-22.3 (t, SnCl$_3$);
(4)	4.87(s, C$_5$H$_5$); 2.25-3.31(m, CH$_2$); 7.16-7.77 (m, C$_6$H$_5$)	81.1(s,C$_5$H$_5$); 28.1(m,CH$_2$); 128.4-133.2 (m, C$_6$H$_5$)	77.4 (s, dppe)	-110.8(t, SnBr$_3$);
(5)	4.77 (s, C$_5$H$_5$); 2.46 (m, CH$_2$); 7.20-7.65 (m,C$_6$H$_5$)	82.4 (s, C$_5$H$_5$); 30.9(s, CH$_2$); 128.3-134.2(m,C$_6$H$_5$)	43.5 (s, dppe)	-31.9(t, SnCl$_3$)
(6)	4.88(s,C$_5$H$_5$); 2.44-2.92(m,CH$_2$); 7.05-7.69 (m, C$_6$H$_5$)	81.1(s,C$_5$H$_5$); 28.2(s, CH$_2$); 128.4-133.2 (m, C$_6$H$_5$)	79.7 (s,dppe)	-344.8 (qt, SnF$_3$);
(7)	4.53(s, C$_5$H$_5$); 7.06-7.35(m, C$_6$H$_5$)	81.3(s,C$_5$H$_5$); 128.2-138.5(m, C$_6$H$_5$)	48.1 (s,PPh$_3$)	-357,0(qt, SnF$_3$);
(8)	4.53 (s, C$_5$H$_5$); 7.06-7.35(m, C$_6$H$_5$)	81.3(s,C$_5$H$_5$); 127.3-138.5(m, C$_6$H$_5$)	45.4(s), 46.3(s), 47.2(s), 48.7(s)^a	-67.6(t), -128,0(dt), -227,2(tt), -357,0(qt)^a
(9)	4.52(s, C$_5$H$_5$); 7.01-7.37(m, C$_6$H$_5$)	81.4(s, C$_5$H$_5$); 127.9-138.6(m, C$_6$H$_5$)	45.6(s), 46.5(s), 47.4(s), 48.4(s)^a	-139.3(t); -159,5(dt); -238,8(tt); -357,9(qt)^a
(10)	4.53(s, C$_5$H$_5$); 7.01-7.37(m,C$_6$H$_5$)	81.3(s, C$_5$H$_5$); 127.1-137.6(m, C$_6$H$_5$)	45.6(s), 45.7(s), 45.8(s)^a	-65.8(t); -89.3(t); -111.6(t)^a
(11)	4.58(s, C$_5$H$_5$); 7.071-7.37(m,C$_6$H$_5$)	81.4(s, C$_5$H$_5$); 128.3-134.6(m, C$_6$H$_5$)	45.3(s)^a	-116 7(t); -139.2(t)^a
(12)	4.88(s, C$_5$H$_5$); 2.17-2.89(m, CH$_2$); 7.20-8.01(m,C$_6$H$_5$)	92.4(s, C$_5$H$_5$); 31.1(s,CH$_2$); 127.9-138.6(m,C$_6$H$_5$)	77.8(s), 78.6(s), 79.2(s), 79.7(s)^a	-24.4(t); -96.9(dt); -206.2(tt); -344,9(qt)
(13)	4.88(s, C$_5$H$_5$); 2.25-3.23(m, CH$_2$); 7.15-7.76(m,C$_6$H$_5$)	81.4(s. C$_5$H$_5$); 27.3(s, CH$_2$); 127 9-138.6(m, C$_5$H$_5$)	78.4(s), 78.6(s), 78.8(s)^a	-51.8(t); -79 0(t)
(14)	4.88(s, C$_5$H$_5$); 2.25-3.36(m, CH$_2$) 7.05-7.70(m,C$_6$H$_5$)	81.4(s, C$_5$H$_5$); 27.1(s,CH$_2$) 127.9-138.6(m, C$_6$H$_5$)	77.5(s)^a	-81.1(t); -110.8(t)^a

s = singlet; t = triplet; dd = doublet of doublets; dt = doublet of triplets; td = triplet of doublets; tt = triplet of triplets; qt = quartet of triplets; m = multiplet. **a** = see text.

3.1 Mössbauer Spectroscopy

3.1.1 Compounds containing the SnX$_3$ moiety (X = Y = F, Cl or Br, complexes 1 to 7)

^{119}Sn-Mössbauer spectrum **(3)** at 77 K is shown in Figure 3 and the corresponding parameters are included in Table 3. Although the complexes in this series, in a strictly chemical sense, are derived from a tin(II) species (SnX$_2$) their isomer shift values, IS, indicate that tin is unmistakably four-covalent [18]. This is explained by the dative character of the Sn→Ru bond, in which a Sn-based non-bonding pair of electrons originally present in SnX$_2$ is shared by the two metal centers upon formation of the metal-metal bond.

From Table 3 it can be seen that the difference between IS and QS values for those complexes for which X = Cl and Br are small but significant, while for those complexes where X = F, there is remarkable change in both values, witness the difference between (1) (IS = 1.98; QS = 1.93) and (7) (IS = 1.53; QS = 2.17). This seems to be due to strong π-backbonding from Ru to Sn which is operative in complexes (6) and (7), for which the increased electronegativity (χ) of the SnF_3 fragment (vs. $SnCl_3$ and $SnBr_3$) causes this fragment to be a better electron-withdrawing agent. While IS values for the X = Cl derivatives fall within the expected range for nearly-sp^3 hybridized tin atoms (~25% s-character) the same variable for X = F indicates an increased importance of the $Sn_{p,d}$ orbitals in the bonding scheme around Sn (and hence reducing the contribution due to Sn_s), exactly the same increase one might expect when π-bonding takes place. The observed QS values indicate a much more pronounced breakdown in the symmetry around the Sn in the SnF_3 fragment in comparison to $SnCl_3$ and $SnBr_3$, possibly indicating an oligomerization of (6) and (7) in the solid state (thus shifting the geometry around Sn from nearly-tetrahedral to distorted octahedral) as well as supporting the existence of π bonding, which also would cause a distortion of FSnF angles.

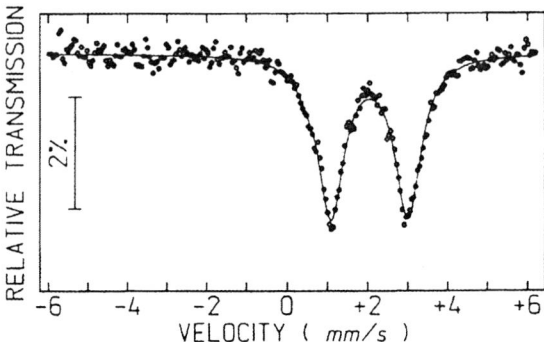

Figure 3 - ^{119}Sn-Mössbauer spectrum at 77 K for $CpRu(dppe)SnCl_3$, (3).

These assertions find support in the ^{119}Sn NMR results. While line position for the $SnBr_3$ and $SnCl_3$ derivatives reflect the overall electronegativity of the groups bonded to Sn (the more electronegative these groups, the higher is the position of ^{119}Sn resonance), tin-resonances when X = F fall in a range typical of a shielded nucleus, implying that electronic density is *donated* to tin, rather than withdrawn, as would be expected on the basis of electronegativity alone.

In fact, by symmetry considerations alone, filled-$Ru_{d_{xy}, d_{xz}, d_{yz}}$ orbitals are suitable to interact via π with empty-$Sn_{d, p}$ orbitals, but they are also involved in bonding with the Cp ring. Only in the case where very electronegative ligands are coordinated to tin, $Sn_{d, p}$ orbitals may compete for the Ru-based electronic density to form a $\pi_{Ru \rightarrow Sn}$ bond.

The linear relationship between IS values and the electronegativity of the ligands around Sn has been demonstrated in several works [18,19] spanning from the early 1970's to date, but seldom applied to the cases in which organometallic fragments are bonded to Sn. The observed values of IS for the $SnCl_3$ derivatives allowed the calculation of the electronegativity of the organometallic fragments. It has been shown [14] that for the species $SnCl_4$, $PhSnCl_3$, Ph_2SnCl_2 and Ph_3SnCl, there is a linear relation between the isomer shift, IS, and the sum of the electronegativity of each of the ligands around Sn, i.e.,

$$IS = -0.9437 \, \Sigma\chi_i + 12.825. \tag{1}$$

where, for complexes (1), (3) and (5), $\Sigma\chi_i = 3\chi_{Cl} + \chi_{[CpRu(phosphine)_2]}$.

Table 3- ^{119}Sn-Mössbauer Parameters at 77 K for the Heterobimetallic Complexes

Compound	IS(mm s^{-1})^a	QS(mm s^{-1})	Γ(mm s^{-1})^b
CpRu(PPh$_3$)$_2$SnCl$_3$	1.98	1.93	0.82
CpRu(PPh$_3$)$_2$SnBr$_3$	2.10	1.79	0.78
CpRu(dppe)SnCl$_3$	1.92	1.94	0.81
CpRu(dppe)SnBr$_3$	2.05	1.80	0.78
{[CpRu(SnCl$_3$)$_2$](μ-dppe)}	1.99	2.03	0.90
CpRu(dppe)SnF$_3$	1.59	2.19	0.79
CpRu(PPh$_3$)$_2$SnF$_3$	1.53	2.17	0.87
CpRu(PPh$_3$)$_2$SnF$_2$Cl	1.92	1.97	0.80
CpRu(PPh$_3$)$_2$SnF$_2$Br	1.78	2.42	0.80
	1.94	1.50	0.80
CpRu(PPh$_3$)$_2$SnCl$_2$Br	2.00	1.92	0.83
CpRu(PPh$_3$)$_2$SnBr$_2$Cl	2.08	1.86	0.83
CpRu(dppe)SnF$_2$Cl	1.68	2.11	0.86
CpRu(dppe)SnCl$_2$Br	1.93	1.69	0.87
	2.00	2.17	0.87
CpRu(dppe)SnBr$_2$Cl	2.03	1.84	0.78

a : Relative to CaSnO$_3$

b : ± 0.01 mm s^{-1} for samples (1) to (8); ± 0.03 mm s^{-1} for samples (9) to (14).

Inserting $\chi_{Cl} = 3.16$ [16] and $IS_{(1)} = 1.98$, $IS_{(3)} = 1.92$, and $IS_{(5)} = 1.99$ the electronegativity (χ_{RM}) of the organometallic fragments, [CpRu(PPh$_3$)$_2$]$^+$, [CpRu(dppe)]$^+$ and [(CpRu)$_2$(μ-dppe)$_2$]$^{2+}$, can be calculated as being 2.01, 2.08, 1.99 respectively. This order of electronegativity values is interesting as it shows the change in the donor ability of dppe as a function of its mode of coordination. The difference of χ_{RM} values between [(CpRu)$_2$(μ-dppe)$_2$]$^{2+}$ and [CpRu(dppe)]$^+$ can be explained by the smaller ring-tension present in the case of μ-dppe than that characteristic of chelating ligands. This difference in tension causes a change in the arrangement of the organic groups around the P atoms, which is observed by the different C-P-C angles in the structures of (1) and {[CpRu(N$_3$)]$_2$(μ-dppe)$_2$} [12]. While the structure of (1) shows an average C-P-C angle of 103.03° the same average for the dimer is of 99.66°. These averages can be used to estimate the s-character of the P-C bonds in dppe, according to the method described by Huheey [16]. For (1) the estimated s-character is 19% while for the dimeric azido compound it is 14%. This small, although not neglectible, difference causes chelating dppe -as in (1) - to be a worse electron donor than its bridging counterpart, complex (5), thus increasing χ_{RM}.

On the other hand, equation (1) could not be used for the case of SnBr$_3$ derivatives. The IS values for (2) and (4) fall outside the range in which this equation was estimated, leading to a high uncertainty of the calculated $\Sigma\chi_i$ values. Also, the use of similar equations based on reported values[17] of IS for R$_{4-n}$SnBr$_n$ compounds did not lead to reasonable estimates of $\Sigma\chi_i$.

The lack of success on estimating $\Sigma\chi_i$ values for the SnBr$_3$ derivatives can be explained based on structural features of (2) and (4). It is known [18,19] that subtle distortions of the coordination environment around Sn can be reflected by considerable changes in IS and QS values. For instance, lengthening of a Sn-X bond in $SnXY_3$ reduces the effective electronegativity of X thus leading to an increase in IS. This lengthening also leads to a breakdown of the symmetry around Sn, disturbing QS values. In the cases of compound (2) and (4), one of the Sn-Br bonds is *ca*. 0.03Å longer than the others. Although this feature is also present in the structure of $SnCl_3$ derivatives, (1) and (3), it is so in a much smaller extent (*ca*.0.02Å), therefore its effect is less pronounced for these compounds. Also, the overall symmetry around Sn decreases from the ideal C_{3v} in (1) and (3), to *nearly* C_{2v} in (2) and (4), with a longer Sn-Br bond. This causes a reduction of the hybridization of Sn from nearly sp^3 to nearly sp^2, causing the observed lower QS values.

3.1.2 Compounds containing the SnX_2Y moiety (X ≠ Y, complexes (8) to (14))

For the case of complexes (8) - (14), ^{119}Sn and ^{31}P NMR spectra (Table 2), show that more than one complex is present in solution. For instance, these spectra show that a sample of (8) is actually comprised of a mixture of four different species, all of which contain the $CpRu(PPh_3)_2^+$ moiety in their structures. This observation is in flagrant disagreement with Mössbauer data (Table 3), which indicate the presence of a single site in the solid state. This is also in disagreement with elemental analysis results, which in turn indicate the expected composition for a single compound, Table 1.

Further analysis of NMR data based on spectral pattern, signal integration and line-position, indicates that (8) is a mixture of $CpRu(PPh_3)_2SnF_3$ (δ ^{119}Sn = -357,0 ppm; ^{31}P = 48.7 ppm), $CpRu(PPh_3)_2SnF_2Cl$ (δ ^{119}Sn = -227,2 ppm; ^{31}P = 47.2 ppm), $CpRu(PPh_3)_2SnFCl_2$ (δ ^{119}Sn = -128,0 ppm; ^{31}P = 46.3ppm) and $CpRu(PPh_3)_2SnCl_3$ (δ ^{119}Sn = -67.6 ppm; ^{31}P = 45.4 ppm), in a 3:2:1:1 ratio, thus explaining the observed elemental analysis results.

That the observed mixture of compounds is retained in the solid-state was confirmed by X-ray crystallography of (14), another case for which a mixture of complexes is formed from the synthesis. Crystallographic empirical formula for this complex is equivalent to a 1:1 mixture of $CpRu(dppe)SnBr_2Cl$ and $CpRu(dppe)SnBr_3$, in agreement with both the ^{119}Sn NMR and elemental analysis results. The same rationale can be applied for the other complexes in this series.

Tin(II) halides are notoriously better trans-halogenating agents than insertion compounds [19]. Therefore it is not surprising that reactions between $CpRu(PR_3)_2Y$ and SnX_2 yielded a mixture of compounds characterized by ligand-redistribution around Sn, rather than the insertion compound $CpRu(PR_3)_2SnX_2Y$ as the sole product.

Halogen quadrupole splitting factors are supposed to be equal or very similar to each other in the X,Y = F, Cl, Br, I series [18]. This being the case, QS value for each component in a given mixture of complex will differ by far less than a typical value of *Γ*, Table 3. Identification of each individual Sn-site by Mössbauer is then precluded, explaining our observations. For this series, ^{119}Sn NMR spectroscopy proved to be a more useful analytical tool.

4. Conclusions

^{119}Sn-Mössbauer spectra of $CpRu(phosphine)_2SnCl_3$ complexes allowed the determination of the electronegativity of several Ru-based organometallic fragments. Electronegativity values correlate well with structural features for those complexes, showing that chelating phosphines are more electronegative than their bridging counterparts.

For the cases of $SnBr_3$ and SnF_3 derivatives, structural features such as Ru→Sn π-backbonding, solid state self-association, or sterically-demanded bond-stretching, caused IS and QS values to fall outside the range in which electronegativity values could be inferred.

The electron-withdrawing ability of the SnF_3 fragment, not shared by the other SnX_2Y moieties, explains the higher catalytic activity of (6) and (7). This fragment causes the development of a higher partial positive charge on the Ru center, increasing its ability to interact with electron-rich organic substrates.

Analysis of Mössbauer spectra for the heteroleptic ($X \neq Y$) complexes was hindered to small differences in the quadrupole splitting abilities among halides.

Acknowledgments

The authors are grateful to the Brazilian agencies FAPEMIG, CNPq and FINEP.

References

[1] P. Smith, "Tin Chemicals", International Tin Research Institute, 1990

[2] T. Blackmore, M.I. Bruce, F.G.A. Stone, J. Chem. Soc. A (1971) 2376

[3] S.G. Davies, J.P. Mcnally, A.J. Smallridge, Adv. Organomet. Chem. 30(1990) 1

[4] M.I. Bruce, M.G. Humphrey, G. Koutsantonis, M.J. Liddell, J. Organomet. Chem. 326(1987) 247

[5] M.I. Bruce, A.G. Swincer, B.J. Thomson, R.C. Wallis, Aust. J. Chem. 33(1980) 2605

[6] G. Domazetis, B. Tarpez, D. Dolphin, B.R. James, J. Chem. Soc. Chem. Comm. (1980) 939

[7] G. Domazetis, B.R. James, B. Tarpez, D. Dolphin, ACS Symp. Ser. 152(1981) 243

[8] (a) G. Stedman, J. Chem. Soc., (1960), 1702, (b) A. Haim, H. Taube, Inorg. Chem. 2(1963) 1199

[9] (a) R.F. Ziolo, Z. Dori J. Am. Chem. Soc. 90 (1968) 6560,
     (b) A.P. Gaughan, K.S. Bowman, Z. Dori Inorg. Chem.11(1972) 601

[10] M. I. Bruce, C. Hameister, A.G. Swincer, R.C. Wallis Inorg. Synth. 21(1982) 78

[11] T. Wilzewski, M. Bochénske, J. F. Bienart, J. Organomet. Chem. 373(1981) 87

[12] A complete report on {[CpRu(N₃)]₂(μ-dppe)} and {[CpRuCl]₂(μ-dppe)} will be published elsewhere.

[13] M.I. Bruce, R.C.F. Gardner, F.G.A. Stone, J. Chem. Soc. Dalton (1976) 81

[14] (a)E.M. Moura, C.M. Rodrigues, D.C. Santos, H.G.L. Siebald and A.Abras, Hyp. Int. (C) 2(1997) 116
     (b)E.V. Marques, W.F. Ribeiro, C.A.L. Filgueiras and A. Abras, Hyp. Int. 96(1995) 259 *and references therein.*

[15] G.M. Sheldrick, SHELXTL-Structure Determination Software Programs , Siemens Analytical X-ray Instruments, Madison, 1990

[16] J. E. Huheey, Inorganic Chemistry - Principles of Structure and Reactivity, Harper International, N. York. 3rd Edition (1983) Chapter 5

[17] P.J. Smith, Organomet. Chem. Rev. A 5(1970) 373

[18] R.S. Drago, Physical Methods in Chemistry, W.B. Saunders, Philadelphia (1977) Chapter 15 *and references therein.*

[19] M.F. Lappert, M.J. McGeary and R.V. Parish, J. Organomet. Chem. 373 (1989) 107

MÖSSBAUER STUDIES OF TIN COMPLEXES AND HETEROBIMETALLIC TIN-TRANSITION METAL COMPOUNDS CONTAINING NITROGEN DONOR LIGAND

Wagner M. Teles[1], Carlos A. L. Filgueiras[2] and Anuar Abras[3*]

[1] *Departamento de Química, ICEx, Universidade Federal de Minas Gerais,*
31270-901 Belo Horizonte , MG (Brazil)
[2] *Departamento de Química Inorgânica, IQ, Universidade Federal do Rio de Janeiro,*
21945-970 Rio de Janeiro, RJ (Brazil)
[3] *Departamento de Física, ICEx, Universidade Federal de Minas Gerais,*
30123-970 Belo Horizonte , MG (Brazil)

A series of tin and heteronuclear tin-transition metal complexes of the nitrogen donor ligand di(2-pyridyl)sulphide, dps, was obtained and its ^{119}Sn Mössbauer studies are here presented. In the complexes [Sn(dps)X$_4$], X = Cl, Br, the Sn atoms are hexacoordinated and bonded to the nitrogens of the ligand. In the heterobimetallic complexes [M(SnCl$_3$)$_2$(dps)], M = Pt, Pd, and [{Pt(SnCl$_3$)(PEt$_3$)Cl}$_2$μ-(dps)] the Sn atoms are tetracoordinated and bonded to the transition metal, as suggested by Mössbauer data.

The measured Mössbauer parameters are also discussed in terms of coordination shifts and showed to be consistent with the heteronuclear NMR and IR spectroscopic data.

Introduction

^{119}Sn Mössbauer spectroscopy has been used extensively to characterise tin-transition metal complexes and obtain information about the electronic environment around the tin atom [1,2]. Several properties such as electronegativity, coordination number and symmetry of tin species can be analysed in terms of isomer shift (δ) and quadrupole splitting (Δ). In fact, in several heterobimetallic compounds containing M-Sn bonds, which are involved in various importants catalytic cycles [3,4] Mössbauer spectroscopy can be useful to understand the nature of this linkage and the structural features of this class of compounds. A series of homo- and heteronuclear tin-transition metal compounds with the di(2-pyridyl)sulphide, dps, was synthesised and characterised by C, H, N and M analyses, and studied by IR and multinuclear NMR spectroscopy [5]. In this work we report the Mössbauer study carried out for these complexes.

The ^{119}Sn spectra were performed in a constant acceleration spectrometer moving a CaSnO$_3$ source at room temperature. The samples were measured at liquid nitrogen temperature. The isomer shift values reported in this work are given with respect to this source. All spectra were computer-fitted assuming Lorentzian lineshapes.

Results and discussion

Figure 1 illustrates some of the Mössbauer spectra, and table 1 gives the Mössbauer data of complexes in study, as well as parameters taken from the literature, used for comparison.

The complexes [Sn(dps)X$_4$], for which X = Cl (**1**) and Br (**2**), show the IR spectra in accordance with tin species bonded to the nitrogen basic sites due to the arise of vibrations atribuited to $\bar{v}_{(Sn-N)}$ and $\bar{v}_{(Sn-X)}$ in the region of low frequencies [5]. The ^{119}Sn Mössbauer spectroscopy presents a consistent lower isomer shift (δ) of (**1**) (0.48 mm/s) and (**2**)

---

* Corresponding author: E-mail: aabras@bach.fisica.ufmg.br

(0.74 mm/s) than the precursors **(13)** (0.82 mm/s) and **(14)** (1.13 mm/s), respectively. The same is true for **(11)** and **(12)**. The reduction of isomer shift values upon complexation has been associated with an increase in the coordination number of tin, which indicates lower s-electron density at the tin nucleus in the hexacoodinated complexes when compared to their precursors. This can be atributed to a greater involviment of the d-orbitals, which now take part in the tin hybridization scheme, thus reducing the weight of s-orbitals in the overall hybridization of the tin [6,7]. Table 1 also shows that in the bromo complex the value of δ is significatively higher than the chloro analog in a agreement with the presence of the more electronegative ligand bounded to the tin centre [2,8]. Similar results have been reported for a great number of other Sn(IV) compounds [6,9,10].

In the case of the heterobimetallic complexes the Mössbauer results reveal that the compounds contain $SnCl_3^-$ ions, but the platinum atoms are covalently bonded to the trichlorostanato species [11]. The isomer shift δ is lower in comparison with other Sn(II) species and the quadrupole splitting is higher, as can be seen in Table 1. This is explained by the character of the Sn→M (M = Pt, Pd) bond, in which the lone pair of electrons originally present on tin is shared by the two metal centres upon formation of the metal-metal bond. However, several spectroscopic studies [12] and theoretical calculations [4] have shown that $SnCl_3$ is a poor σ- donor and a good π-acceptor. Thus, in tin-transition metal complexes the Mössbauer parameters can be tentatively interpreted in terms of synergic (σ + π) interactions between tin and the transition metal. The π back-donation from the filled d orbitals on M to the empty $5p_z$ (or 5d) orbitals on tin causes a decrease in the effective 5s-electron density of the tin by a shielding effect, and therefore the isomer shifts follows the same trend. Of course, if the contribution of π back-donation is not pronounced this influence will be not affect the Mössbauer parameters.

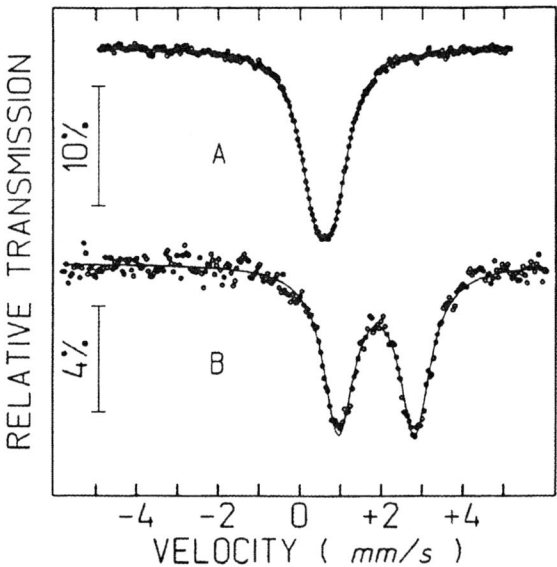

Figure 1. Mössbauer Spectra at liquid nitrogen temperature of:
A) [Sn(dps)Cl₄] **(1)** and B) [{Pt(SnCl₃)(PEt₃)Cl}₂μ-(dps)] **(5)**

In order to analyse the Mössbauer parameters in our heterobimetallic complexes, we used the coordination shifts formalism [13]. In this context, the reduction in population of the 5s and 5p orbitals of tin, Sc and Pc, due to coordination to the transition metal centre is related to the Mössbauer parameters as follows.

[119]Sn Mössbauer spectra are characterised by two main parameters, the isomer shift (δ) and the quadupole splitting (Δ). For a given isotope, δ depends on the electron density at the nucleus. Therefore the difference in δ between two compounds is proportional to the difference in electron density which, in turn, is a measure of the difference in population of the valence-shell orbitals. Thus, the relevant difference is that between the free ligand and its complexes, i.e., $\delta c = \delta_{cpx} - \delta_{lig}$, where δc is the coordination isomer shift and cpx

Table 1. [119]Sn Mössbauer parameters measured at liquid nitrogen temperature.

Compound	δ(mm/s)	Δ(mm/s)	Sc	Pc	Sc + Pc
(1) [Sn(dps)Cl$_4$]	0.48(1)	0.47(2)			
(2) [Sn(dps)Br$_4$]	0.74(1)	0.52(2)			
(3) [Pt(SnCl$_3$)$_2$(dps)]	1.67(1)	1.85(1)	0.68	-0.21	0.47
(4) [Pd(SnCl$_3$)$_2$(dps)]	1.85(1)	1.76(1)	0.62	-0.19	0.43
(5) [{Pt(SnCl$_3$)(PEt$_3$)Cl}$_2$$\mu$-(dps)]	1.75(1)	1.88(1)	0.65	-0.22	0.43
(6) trans-[Pt(SnCl$_3$)Cl(PPh$_3$)$_2$][a]	1.84	2.07	0.61	-0.27	0.34
(7) trans-[Pt(SnCl$_3$)H(PPh$_3$)$_2$][a]	1.94	1.98	0.57	-0.25	0.32
(8) cis-[Pt(SnCl$_3$)$_2$Cl$_2$](Et$_4$N)$_2$[a]	1.61	2.01	0.70	-0.25	0.45
(9) cis-[Pd(SnCl$_3$)$_2$Cl$_2$](Me$_4$N)$_2$[a]	1.52	2.21	0.73	-0.30	0.43
(10) [Pt(SnCl$_3$)Cl(dppp)][b]	1.95	1.97	0.58	-0.24	0.34
(11) [Sn(2,2'-py$_2$CO)Cl$_4$][c]	0.42(1)	0.39(2)			
(12) [Sn(2,2'-py$_2$CO)Br$_4$][c]	0.54(1)	0.55(2)			
(13) SnCl$_4$[c]	0.82(1)	0			
(14) SnBr$_4$[c]	1.13(1)	0			
(15) SnCl$_2$[d]	4.06(2)	0.66(4)			
(16) CsSnCl$_3$[a]	3.40	1.22			

[a] Ref.[13]; [b] Ref.[11], dppp = 1,3-bis(diphenylphosphino)propane
[c] Ref.[8], 2,2'-py$_2$CO = 2,2'-bis(pyridyl)ketone; [d] Ref. [17]

and lig stand for complex and ligand, respectively. Since only s-electrons have finite probability of occurring at the nucleus, the contribution of the other electrons to δ is only indirect and very small. Thus, δc measures mainly the change in valence-shell s-electron density on the tin atom.

On the other hand, Δ arises from the interaction of quadrupolar nuclei with an electric-field gradient (efg) produced by an asymmetric distribution of charge about the nucleus. The main contribution to efg comes from an imbalance in the population of the valence-shell p-orbitals. In the absence of polarization effects, the set of three p orbitals have spherical symmetry. Also, the contribution from charges beyond the valence-shell is negligible because of the inverse cubic dependence of efg on the distance from the nucleus. By a similar procedure, the coordination shift in the quadrupole splitting can be defined as $\Delta c = \Delta_{cpx} - \Delta_{lig}$.

A better interpretation of the data is done using a calibration scale of δ and Δ in terms of electron configurations. To a good approximation, δ depends directly on the valence-shell s population and indiretly (by shielding) on the p population:

$$\delta = an_s + bn_p \qquad (1)$$

and Δ depends on the p-imbalance:

$$\Delta = cUp = c[n_z - 1/2(n_x + n_y)] \qquad (2)$$

where n_s, n_x, n_y and n_z are the population of the 5s, $5p_x$, $5p_y$ and $5p_z$ orbitals. The best available values for the coefficients are a = -2.7 mm/(s × e), b = 0.15 mm/(s × e) and c = 4.0 mm/(s × e), when δ is given with respect to SnO_2 [13].

In terms of the coordination shifts, δc and Δc, equations (1) and (2) become δc = aSc + bPc and Δc = cPc, respectively, where Pc and Sc are the number of p and s electrons lost to the acceptor. Solving the last two equations for Pc and Sc we get using the coefficients: Sc = -(δc - 0.15Pc)/2.7 (mm/s) and Pc = -Δc/4.0 (mm/s). The minus signs in these expressions indicate that we are calculating the electron density removed from the tin atom upon coordination. To obtain the coordination shifts, δc and Δc, using the measured δ_{cpx} and Δ_{cpx} of a given transition metal complex containing the ligand $SnCl_3$, we took the compound [Et$_2$OH][SnCl$_3$] as a reference, for which δ_{lig} = 3.54 mm/s and Δ_{lig} = 1.00 mm/s [13]. Thus, the coordination shifts of a complex having isomer shift δ and quadrupole splitting Δ are given by δc = (δ - 3.54) (mm/s) and Δc = (Δ - 1.00) (mm/s), respectively.

Table 1 shows the calculated Sc and Pc, as well as (Sc + Pc) which is always a net loss of electron density by tin to the transition metal. Thus (Sc + Pc) is always positive and measures an Sn→M charge donation. In our compounds the (Sc + Pc) values are higher than for other analogs, indicating a considerable donation from Sn to Pt. Comparing the (Sc + Pc) values of compounds [M(SnCl$_3$)$_2$(dps)], for which   M= Pt **(3)** and Pd **(4)**, one sees that the tin charge donation increases when the principal quantum number increases, i.e., from Pd (4d) to Pt (5d). The pair **(8)** and **(9)** is in agreement with this observation.

The (Sc + Pc) parameters are sensitive to the nature of the bond in a trans-position relative to the tin atoms. In fact, in trans-[Pt(SnCl$_3$)H(PPh$_3$)$_2$] **(7)** one sees that the Sn→M donation occurs less than in trans-[Pt(SnCl$_3$)Cl(PPh$_3$)$_2$] **(6)**, due to the greater trans-influence of hydride compared to chloride. The same behaviour is evidenced when   [Pt(SnCl$_3$)$_2$(dps)] **(3)** and [Pt(SnCl$_3$)Cl(dppp)] **(10)** are compared, in agreement with the trans-influence series chloride < pyridine-donor < P-donor < hydride. Despite the facts discussed above, if we compare cis-[M(SnCl$_3$)$_2$Cl$_2$]$^{2-}$, for which M = Pt **(8)** and Pd **(9)** with [M(SnCl$_3$)$_2$(dps)], for which M = Pt **(3)** and Pd **(4)**, we would expect a less Sn→M donation, but the results show that this is not the case. This anomaly   can be tentatively associated with the fact that compounds **(8)** and **(9)**   are anionic.

An interesting situation can be seen regarding the complex [{Pt(SnCl$_3$)Cl(PEt$_3$)}$_2$μ-(dps)] **(5)**. This complex has a higher δ than complex [Pt(SnCl$_3$)$_2$(dps)] **(3)**, indicating a higher s density on the Sn atom. This fact can be extended when we analyse the ^{119}Sn NMR data relative to ^1J(^{195}Pt-^{119}Sn) coupling, in which the magnitude of ^1J(^{195}Pt-^{119}Sn) is proportional to the product of the s-electron spin density at the two nuclei, among other factors in the Fermi contact expression [14]. As expected the ^1J(^{195}Pt-^{119}Sn) coupling in complex **(5)** is higher than in **(3)**, which reveals a stronger Pt-Sn interaction in agreement with the fact that chloride has a lower trans-influence than N-pyridine. However, the same analysis does not apply if we compare **(5)** and **(7)**, since the latter presents a higher value of δ and a lower ^1J(^{195}Pt-^{119}Sn) than **(5)** [15] due the high trans-influence of the hydride ligand. This seemingly contradictory result can be tentatively explained if we observe that in **(5)** the (Sc + Pc) term is higher than in **(7)**, implicating a more efective Pt-Sn bond in **(5)** in accordance with the ^{119}Sn NMR results. In addition, if we consider that Pt-Sn π back-donation is favored in **(5)** and therefore causes a decrease in 5s-electron density on tin, the value of δ in complex **(5)** is not so puzzling.

The ^{119}Sn Mössbauer spectra show that in the [M(SnCl$_3$)$_2$(dps)] complexes we have two equivalent tin atoms, which is corroborated by the ^{119}Sn NMR results. The values of δ are consistent with those for other complexes synthesised in our laboratory containing tetrahedral tin species bonded to electron rich (d^8) transition-metals [16]. On the other hand, the Mössbauer spectrum of complex [{Pt(SnCl$_3$)(PEt$_3$)Cl}$_2$μ-(dps)], presented in Figure 1.B, shows only one Sn site. This is not the case, however, when the tin species is present in solution as indicated by the ^{119}Sn NMR spectra, due to fluxional behaviour of this complex [5].

These results show that Mössbauer spectroscopy provides a useful technique to study the nature of the Pt-Sn bond in correlation with other spectroscopic methods. However, care is needed to analyse the Mössbauer parameters due to the complexity of the heterobimetallic systems.

Acknowledgement

The authors are grateful to partial financial support from CNPq, FAPEMIG and FINEP. W. M. T. also thanks to Dr. W. R. Rocha for helpful discussions.

References

[1] R. J. Dickinson, R. V. Parish, J. Rowbotham, A. R. Manning and Paul Hackett, J. Chem.
    Soc. Dalton (1975) 424.
[2] E. V. Marques, W. F. Ribeiro, C. A. L. Filgueiras and Anuar Abras, Hyp. Int. 96 (1995) 259.
[3] R. Bardi, A. M. Piazzesi, G. Cavinato, P. Cavoli and L. Toniolo, J. Organomet. Chem. 224 (1982) 407.
[4] W. R. Rocha and W. B. De Almeida, Int. J. of Quantum Chem. 65 (1997) 643.
[5] W. M. Teles, N. G. Fernandes, Anuar Abras and C. A. L. Filgueiras, submited for publication in Transition Metal Chem.
[6] R. S. Randall, R. W. J. Wedd and R. J. Sams, J. Organomet. Chem. 30 (1971) C19.
[7] D. Cunningham, M. Little and K. McLoughlin, J. Organomet. Chem.165 (1979) 287.
[8] W. M. Teles, L. R. Allain, C. A. L. Filgueiras and Anuar Abras, Hyp. Int. 83 (1994) 175.
[9] R. S. Barbieri, H. O. Beraldo, C. A. L. Filgueiras, A. Abras, J. F. Nixon and P. B. Hitchock, Inorg. Chim. Acta 206 (1993) 169.
[10] Anuar Abras, G. F. De Sousa and C. A. L. Filgueiras Hyp. Int. 90 (1994) 459.
[11] Tamás Kégl, László Kollár, Gábor Szalontai, Erno Kuzmann, Attila Vértes, J. Organomet. Chem. 507 (1996) 75.
[12] (a) G. W. Parshall, J. Am. Chem. Soc. 88 (1966) 704.
    (b) A. Albinati, U. Von Guten, P. S. Pregosin and H. J. Ruegg, J. Organomet. Chem. 295 (1985) 239.
[13] R. V. Parish, Coordination Chemistry Reviews 42 (1982) 1.
[14] K. H. A. Ostoja Starzewski and P. S. Pregosin in "Catalytic Aspects of Metal Phosphine Complexes", Advances in Chemistry Series 196 (1982) 23.
[15] A. Albinati, R. Naegeli, K. H. A. Ostoja Starzewski, P. S. Pregosin and H. Rüegger, Inorg. Chim. Acta 76 (1983) L231.
[16] G. M. De Lima, C. A. L. Filgueiras and Anuar Abras, Hyp. Int. 83(1994) 183.
[17] T. Birchall, P. A. W. Dean and R. J. Gillespie, J. Chem. Soc. A (1971) 1777.

IRON OXIDES OF A MAGNETIC SOIL DERIVED FROM DOLOMITIC ITABIRITE

C. S. de Moura[1], J. D. Fabris[1], W. da N. Mussel[1], C. A. Rosière[2]

[1]Departamento de Química, UFMG, 31270-901 Belo Horizonte, MG, Brazil
[2]Departamento de Geologia, UFMG, 31270-901 Belo Horizonte, MG, Brazil

Three soil samples were collected from a magnetic Inceptsol derived from dolomitic itabirite. The pedon was located in iron mining in the State of Minas Gerais, Brazil. Magnetic concentrates from soil sand fractions of samples from A and B horizons were separated by picking up magnetic particles with a hand magnet. Chemical analysis of these fractions indicates a composition of about (in mass%): 93.3 – 93.5 % Fe_2O_3; 2.7 - 2.8 % Al_2O_3; 0.8 - 0.9% SiO_2; 0.53 % TiO_2; 0.04 - 0.06 % CaO and 0.02 – 0.04 % MgO. X-ray and Mössbauer analysis reveal that hematite (determined lattice parameters of hexagonal cell, $a = 0.499 \pm 0.001$ nm; $c = 1.366 \pm 0.004$ nm) and maghemite (cubic cell, $a_0 = 0.8314 \pm 0.0005$ nm) are the main iron oxides of the magnetic extract from the sand fraction of the A horizon.

1. Introduction

Soils are magnetic mainly due to the presence of ferrimagnetic iron oxides with spinel structure, namely maghemite (γFe_2O_3) and magnetite (Fe_3O_4). The practical importance of magnetic soils from mafic lithology in Brazil has been reviewed in several relatively recent papers [1] [2] [3]. Those derived from non-mafic (e.g. itabirite, lateritic and hematite-rich rocks, steatite or ferroan dolomite) parent materials are distinguishable in major aspects of their geology, iron oxide mineralogy and mechanisms of pedogenesis. In Brazil, soils forming on itabirite, a metamorphic rock that contains essentially hematite and quartz, do occur in the heterogeneous and very old (Precambrian) geodomain of Quadrilátero Ferrífero (Minas Gerais State), an area that provides economically exploitable iron ore deposits. Those soils specifically have poor natural fertility and therefore limited agriculture potential. Hematite is by far the dominant iron oxide in the area, but lithogenic maghemite has been also found [4] in soils forming on itabirite. The way iron oxides are formed and transformed during pedogenesis in this lithology is not well known. For instance, Allan et al [5] described a soil with magnetization higher than that of the parent itabirite, which was interpreted as being due to the pedogenic transformation of hematite to maghemite. Moukarika et al. [6] identified the maghemite of a magnetic soil in Quadrilátero Ferrífero as being neoformed from iron released from the lattice of the parent ferroan dolomite. The main purpose of the present work was to study, in some detail, samples from a profile of a magnetic pedon forming on dolomitic itabirite in the Quadrilátero Ferrífero, in an attempt to characterize their iron oxides, and by so doing, to add data which could help forming a more general picture about the mineralogy and mineral genesis in that lithodomain.

2. Material and Methods

The samples were collected in three depths, at top (A horizon), between 50 – 60 cm (B horizon) and between 110 – 140 cm (C horizon) deep, in a profile of a magnetic Inceptsol found in an iron mining area (geographical coordinates 20° 09' S 43° 58' W), in Quadrilátero Ferrífero, Minas Gerais State, Brazil.

All samples were first broken up by hand, left to dry in air and sieved, to obtain the fine earth (mean diameter of particles, $\phi < 2$ mm). The particle size distribution of the fine earth was determined by dispersing the sample with a solution prepared by taking 50 mL NH_4OH 1:1 into a final volume of 500 mL, under stirring in a mixer. The suspension was then sieved, to separate the sand fraction (ϕ = 2 mm - 0.05 mm). The silt (ϕ = 0.05 mm - 0.002 mm) and the clay (ϕ < 0.002 mm) fractions were separated by centrifuging the remaining at 1000 rpm for 3 minutes [7] [8]. Magnetic separation was performed on the sand fraction of samples from A and B horizons, by picking up magnetic particles with a hand magnet. Chemical analysis [9] of Si, Ti, Al, Ca, Mg, Co, Cr, Ni, Ba, and low contents of Fe was performed firstly by fusing the sample in a Na_2CO_3 + $Na_2B_4O_7$ and reading in an ICP Spectro Modula equipment. Higher contents of Fe were determined by volumetric method titration with $K_2Cr_2O_7$. Ferrous iron was brought to solution by attacking the sample with HCl under inert atmosphere of CO_2, followed by volumetric determination with $K_2Cr_2O_7$. Potassium was analysed by attacking the sample with a mixture of HNO_3 + HCl in a microwave oven and then reading in an ICP Spectro FM03 equipment. Loss on ignition was gravimetrically determined by burning the sample in a furnace at 1000 °C for 2 hours. The X-ray diffraction patterns were obtained with a Rigaku Geigerflex diffractometer equipped with a graphite diffracted beam monochromator, using CuK_α radiation. NaCl was used as internal standard. Room-temperature Mössbauer spectra were recorded in a conventional constant-acceleration transmission spectrometer and a Co^{57}/Rh source. Magnetization measurements were performed in a portable magnetometer [10]. Results of both, chemical composition and saturation magnetization, are presented in Table 1.

Table 1 - Magnetization and chemical composition of samples.

Horizon	Particle size (subfraction[*])	Mass%	Magnetization σ/J T^{-1} kg^{-1}	Fe_2O_3	TiO_2	Al_2O_3	CaO	MnO	MgO	SiO_2	LOI[**]
							Mass%				
A	Soil (WS)		5.35	ND	ND	ND	ND	ND	ND	ND	ND
	Sand (RF)	31	4.37	66.0	0.92	10.0	0.18	0.68	0.20	9.9	11.3
	Sand (MS)	13	25.86	93.3	0.53	2.78	0.06	0.52	0.02	0.9	ND
	Silt (WS)	45	5.34	41.2	1.79	20.5	0.42	0.43	1.0	14.8	20.3
	Clay (WS)	9	3.41	ND	ND	ND	ND	ND	ND	ND	ND
B	Soil (WS)		4.02	ND	ND	ND	ND	ND	ND	ND	ND
	Sand (RF)	25	1.92	66.8	0.78	12.8	0.08	1.38	0.56	10.0	6.9
	Sand (MS)	8	31.35	93.5	0.53	2.74	0.04	0.45	0.03	0.82	ND
	Silt (WS)	58	4.80	45.6	1.84	20.0	0.18	0.37	1.5	16.7	13.1
	Clay (WS)	7	2.54	42.4	0.95	26.1	0.06	0.23	0.30	10.3	19.5
C	Sand (WS)	10	ND	42.1	0.4	25.9	0.08	0.09	0.56	18.3	10.9
	Silt (WS)	81	ND	25.6	1.17	22.5	0.04	0.08	5.3	35.7	8.4
	Clay (WS)	4	ND	53.5	0.82	12.9	0.1	0.19	0.51	13.3	17.3
	Rock		ND	83.2	0.30	4.93	0.01	0.13	0.02	7.08	3.3

WS = Whole Sample; MS = Magnetic Separate and RF = Remaining Fraction after magnetic separation.
** Loss on ignition. ND = not determined Coefficient of variation, taken as CV = 100 x (standard deviation)/mean, are within the interval 0 ≤ CV < 3 % for all values, basing on three independent determinations.

3.   Results and Discussion

Powder X-ray diffraction patterns (figures not shown) reveal the presence of gibbsite, hematite, quartz and a spinel phase, which, as will be shown later on in this work, is actually maghemite, in the sand and silt fractions from both A and B horizons. Corresponding patterns for the magnetic separate of the sand fractions from A and B horizons (Figure 1, (a) and (b), respectively) evidenced the characteristic reflections of both maghemite and hematite. The lattice parameters of these two oxides were determined by simultaneously least square fitting the main reflections 311, 220, 400, 333, 440, for maghemite and 104, 024, 110, 300, 214, 113, 116, for hematite. Results presented (Table 2) confirm that the cell edge for the cubic spinel phase corresponds more characteristically to that of maghemite though parameters, that range from a_0 = 0.827(3) to 0.836(5) nm, are somewhat variable from sample to sample, but tend to be lower than that of the reported value for the stoichiometric oxide (a_0 = 0.8350 nm, JCPDS [11] card # 24-81).

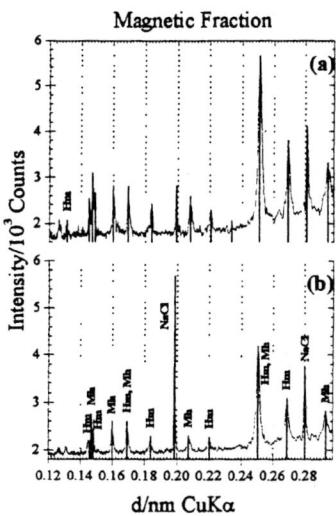

Figure 1 - Powder X-ray patterns for the magnetic separate from sand fraction from (a) A- and (b) B-horizon. Hm = hematite; Mh = maghemite. NaCl = internal standard.

For hematite, determined lattice parameters are: a = 0.499(1) − 0.504(1) nm and c = 1.353(5) − 1.391(5) nm. The a-values lie somewhat below the expected value for stoichiometric hematite, according to reported data in JCPDS, a = 0.5038 nm (card # 24-72; c = 1.3772 nm), probably due to some isomorphic replacement of Fe^{3+} by a smaller cation, such as Al^{3+}. Roughly, a values for both maghemite and hematite tend to decrease from the top to the bottom of the profile. The structural aluminium content in hematite can be estimated from the lattice contraction, in several numerical correlations. For instance, Kämpf and Schwertmann [12] recently proposed the linear correlation Al/(mol mol^{-1}) = 35.46 − 70.402 a/nm, between isomorphic Al-content extractable with citrate-bicarbonate-dithionite, and a–cell parameter for hematites from Brazilian soils. Basing on this equation, the Al content of the presently studied hematites varies from nearly zero (remaining fraction after magnetic separation from sand of A-horizon, a = 0.504(1) nm) to 0.329 mol mol^{-1} (magnetic separate from sand of A-horizon, a = 0.499(1) nm).

Values of saturation magnetization (σ, Table 1) along with Mössbauer data of the magnetic separates suggest that maghemite is the only iron-rich spinel present.

Table 2 - Determined unit cell parameters for the iron maghemite and hematite.

Soil fraction (horizon) and subfraction*	Mineral	Determined parameters		Unit cell volume
		a/nm	c/nm	V/nm^3
Sand (A) RF	maghemite	0.836(5)	-	583(10)
	hematite	0.504(1)	1.391(5)	306(1)
Sand (A) MS	maghemite	0.8314(5)	-	574.8(9)
	hematite	0.499(1)	1.365(7)	292(1)
Silt (A) WS	maghemite	0.8305(7)	-	573(1)
	hematite	0.5031(7)	1.372(5)	300.9(9)
Sand (B) RF	maghemite	0.827(3)	-	567 (6)
	hematite	0.5023(9)	1.364(6)	298(1)
Sand (B) MS	maghemite	0.827(1)	-	567 (3)
	hematite	0.502(1)	1.357(9)	297 (2)
Silt (B) WS	maghemite	0.827(3)	-	572(5)
	hematite	0.4992(8)	1.364(5)	294.3(8)
Sand (C) WS	hematite	0.501(1)	1.353(5)	295(1)
Silt (C) WS	hematite	0.4995(3)	1.373(3)	296.6(6)

WS = Whole Sample; MS = Magnetic Separate and RF = Remaining Fraction after magnetic separation.

The magnetic separate from the sand fraction of B horizon gives a magnetization which is about 20 % above the value measured for the corresponding sample from A horizon, but this probably reflects the difference in the efficiency of the magnetic separation procedure, rather than an intrinsic property of the maghemite. Data in Table 1 show that σ tends to decrease from A to B horizons, either for the whole soil and for all particle-size fractions. It is quite difficult to obtain a reliable quantification of maghemite in presence of hematite, by zero-field Mössbauer spectroscopy, even from the magnetic separate subfraction (Figure 2), as uncertainties on the estimation of hyperfine parameters and subspectral areas are expected to be high. In the present case, two subspectra of maghemite were used to fit Mössbauer data of magnetically separated samples, from sand fractions of A and B horizons (Figure 2). Alternatively, any attempt to fit spectra by including a magnetic field distribution for maghemite, along with a fixed subspectrum for hematite does not improve significantly the results. As an example, Figure 3a; shows such a fitting for the spectrum obtained for the magnetic separate portion of the sand fraction from A horizon. Averaged hyperfine parameters taken from the distribution represented in Figure 3b are: isomer shift relative to αFe, δ = 0.34 mm s^{-1} and quadrupole shift $2\varepsilon_Q$ = -0.05. The hyperfine field distribution profile shows a broad and asymmetric main peak, spanning from ca. 45.3 tesla to 50.4 tesla, with B_{hf} ~ 49 tesla at maximum probability, and a shoulder around ~ 47 tesla. These values from hyperfine field distribution are roughly the same obtained from conventional fitting, with single hyperfine field method. Corresponding fitted parameters are presented in presented in Table 3. In view of this, relative subspectral areas from Figure 3, are roughly taken as quantitative values for the relative iron oxide occurrence in all samples, discarding any possible differences in the f-factors for the iron sites in theses minerals. If so, the corresponding saturation magnetization σ = 26 J T^{-1} kg^{-1} and 31 J T^{-1} kg^{-1} of samples provide some basis to estimate the magnetization of this soil-maghemite, which may lie between 40 J T^{-1} kg^{-1} and 54 J T^{-1} kg^{-1}.

This result is consistent with compiled values of magnetization for maghemite in several

Brazilian soils, though derived from mafic lithology [3].

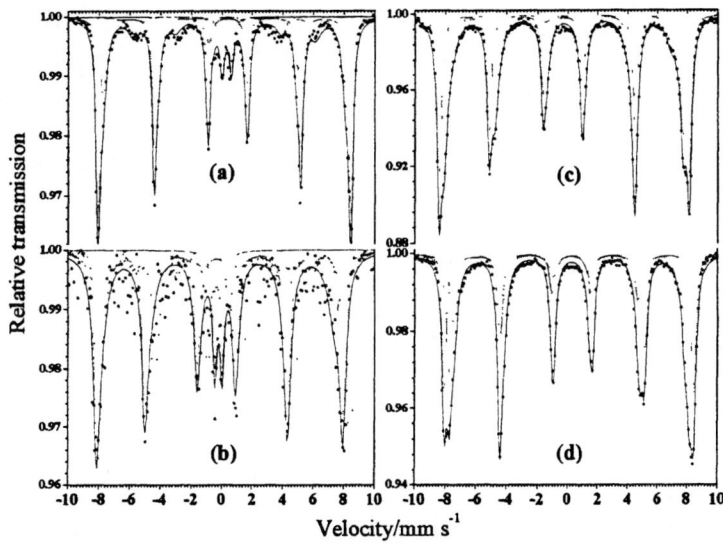

Figure 2 - Room temperature Mössbauer spectra for the portion remained from the magnetic separation, and for the magnetic separate itself from the sand fraction of (a) A- and (b) B-horizon.

Table3 - Mössbauer parameters at RT of the whole samples and of the magnetic separates of the sand fraction. δ = isomer shift relative to αFe; ΔE_Q = quadrupole splitting; $2\varepsilon_Q$ = quadrupole shift; B_{hf} = hyperfine field, RA = relative area of the subspectrum.

Soil fraction (horizon) and subfraction[*]	Subspectrum	δ/mm s^{-1}	B_{hf}/T	ΔE_Q, $2\varepsilon_Q$/mm s^{-1}	RA/%
Sand (A) RF	maghemite	0.36	48.41	0.00	5
	hematite	0.39	49.16	-0.22	20
	hematite	0.36	51.2	-0.19	56
	goethite	0.36	35.09	-0.25	13
	doublet	0.49	-	0.6	6
Sand (A) MS	maghemite	0.32	47.63	-0.05	28
	maghemite	0.31	49.5	-0.01	36
	hematite	0.36	51.35	-0.17	36
Sand (B) RF	maghemite	0.36	47.7	0.00	42
	hematite	0.34	50.01	-0.20	48
	doublet	0.33	-	0.50	10
Sand (B) MS	maghemite	0.44	45.2	0.00	45
	maghemite	0.30	49.21	0.00	12
	hematite	0.30	51.47	-0.19	42

* MS = Magnetic Separate and RF = Remaining Fraction after magnetic separation.

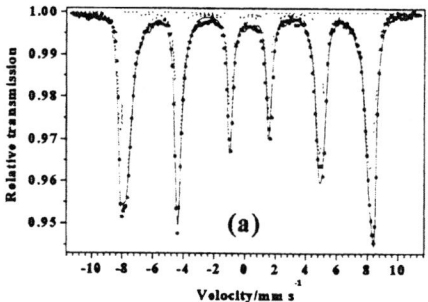 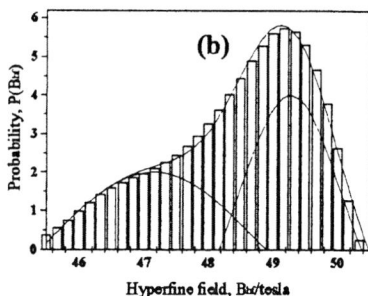

Figure 3 – (a) Room temperature Mössbauer spectrum of the magnetic separate of the sand fraction from A horizon, fitted with hyperfine field distribution for maghemite, and a fixed magnetic subspectrum for hematite and (b) the corresponding hyperfine field distribution profile.

4.   Conclusions

Chemical and mineralogical analyses of samples from a soil profile derived from itabirite revealed that (i) hematite and maghemite are the main iron oxides present; (ii) lattice parameters of hematite may indicate that isomorphical substitution for iron does occur; (iii) the soil magnetization decreases from the top to the bottom of the profile and (iv) maghemite has an intrinsic saturation magnetization between 40 J T^{-1} kg^{-1} and 54 J T^{-1}, lying within the reported upper and lower values for this quantity observed for the maghemite phases in Brazilian soils derived from Brazilian soils though derived from mafic lithology.

Acknowledgements

Work supported by FAPEMIG, CNPq and FINEP (Brazil).

References

[1] M. Resende, D.P. Santana, D.P. Franzmeier, J.M.D. Coey, in: Proc. of the Eighth International Soil Classification Workshop, F.H. Beinroth, M.N. Camargo, H. Eswaran, H, eds. (EMBRAPA, Rio de Janeiro, 1988) Part 1: Papers, p. 78.

[2] J.D. Fabris, M.F. de Jesus Filho, J.M.D. Coey, W. da N. Mussel, A. T. Goulart, Hyp. Int. 110 (1997) 23.

[3] J.D. Fabris, J.M.D. Coey, W. da N. Mussel, Hyp. Int. 113 (1998) 249.

[4] N. Curi, D.P. Franzmeier, Soil Sci. Soc. Am. J. 51 (1987) 153.

[5] J.E.M. Allan, J.M.D. Coey, M. Resende, J.D. Fabris, Phys. Chem. Min. 15 (1988) 470.

[6] A. Moukarika, F. O'Brien, J.M.D. Coey, Geophys. Res. Let. 18 (1991) 2043.

[7] Empresa Brasileira de Pesquisa Agropecuária, in: Manual de métodos de análise de solo (EMBRAPA, Rio de Janeiro, 1997) 212 p. In Portuguese.

[8] M.L. Jackson, Soil chemical analysis: advanced course, 3rd ed. (published by the author,

Madison, Wisconsin, 1969) 894 p.

[9] Jeffery P.G., Hutchison D., Chemical methods of rock analysis. (Pergamon Press, Oxford, 1981). 379p.

[10] J. M. D. Coey, O. Cugat, J. McCauley, J. D. Fabris, Revista de Física Aplicada e Instrumentação 7 (1992) 25.

[11] Joint Committee on Powder Diffraction Standards, 1980. Mineral Powder Diffraction Files Data Book, Swarthmore, Pensilvania.

[12] N. Kämpf, U. Schwertmann, Rev. Bras. Ci. Solo 22 (1998) 209.

STRUCTURAL ASPECTS IN TeGaSn AMORPHOUS ALLOYS INVESTIGATED BY ^{119}Sn MÖSSBAUER SPECTROMETRY

M. Fontana[1], B. Arcondo[1] and J.M. Greneche[2].

1.   Departamento de Física, Facultad de Ingeniería, Universidad de Buenos Aires,
Paseo Colón 850, (1063) Buenos Aires, Argentina.
2.   Laboratoire de Physique de l'Etat Condensé, UPRES A CNRS 6087, Université du Maine,
72085 Le Mans Cedex, France.

The $Ga_x Te_{(100-x)}$ system, when quenched from the melt with the piston and anvil technique, exhibits a small glass forming composition range for $20 \leq x \lesssim 23$ atomic %. Addition of Sn increases the glass forming ability range and favors the amorphization of GaTe alloys. Cast and rapid quenched samples of ternary system GaTeSn were studied by means of X-ray diffraction and ^{119}Sn Mössbauer spectrometry at 77 K. In the crystalline samples the divalent Sn atoms are always surrounded by Te in octahedrally coordinated environments. It is observed that the addition of Sn, above a threshold concentration, stabilizes the high temperature phase Ga_2Te_5 at room temperature. The Mössbauer spectra obtained on amorphous samples were analyzed with either two quadrupolar components with slightly broadened lines or quadrupolar splitting and isomer shift distributions linearly correlated. Both descriptions suggest two kinds of environments at Sn sites: one ascribed to the octahedral coordination, the other one to Sn surrounded by Te with tetrahedral coordination. The present results are also compared with other TeSn based amorphous alloys.

1. Introduction

One of the characteristics of the $Ga_x Te_{100-x}$ system is the existence of a glass forming range around x = 20 atomic %. Amorphous $Ga_{20}Te_{80}$ alloys have been obtained by rapid quenching from the melt [1,2] and Luo et al. have reported a glass forming composition range x from 10 to 30 [3] whereas amorphous samples were prepared by vapor deposition with x comprised from 20 to 60 [4].

Two equilibrium phases are reported in the Ga-Te phase diagram at room temperature for x from 0 to 40 at. %: cubic Ga_2Te_3 and hexagonal Te. Ga_2Te_3 crystals doped with Sn were studied by Mössbauer spectrometry and the spectrum at 80 K consists in a single line with IS = 3.73 mm/s. IS is given relative to SnO_2 [5]. Besides, an other phase, with stoichiometry Ga_2Te_5, is evidenced at temperatures from 679K to 761K [6].

In the Sn-Te system, a stoichiometric phase, SnTe, with a narrow range of homogeneity around 50 at. % Te is formed [7]. A second-order phase transition occurs in SnTe at low temperatures. The critical temperature of this transformation is strongly dependent on composition with values comprised between 24 K and 145 K. Hatta et al [8] and Valassiades et al [9] suggested that this transformation occurs from the cubic NaCl-type structure to rhombohedral or orthorhombic structure. The Mössbauer spectrum of this phase using a $Ba^{119}SnO_3$ source shows a single line with an isomer shift IS = (3.48 ± 0.03) mm/s at room temperature, and the appearance of a small quadrupolar splitting QS = (0.31 ± 0.04) mm/s with IS = (3.57 ± 0.03) mm/s at 80 K [10].

1.1 Glass forming composition range and crystallization of amorphous GaTeSn alloys.

Fontana et al [11] reported that amorphous samples $Ga_x Te_{100-x}$ have been obtained in the range x=20-23 by rapid solidification from the melt with a piston and anvil device. An addition of Sn to the binary system GaTe favored the amorphization, increasing the glass forming range. For example, a full amorphization was obtained with the addition of 1.5 at. % Sn to $Ga_{13}Te_{87}$. X-Ray Diffraction (**XRD**) and Differential Scanning Calorimetric (**DSC**) have been used to study the crystallization kinetics of amorphous alloys GaTeSn. Two main exothermic transformations are reported in the $Ga_{20}Te_{80}$ crystallization. They correspond to a primary crystallization of Te and a secondary crystallization of the high temperature phase Ga_2Te_5. Provided that just one exothermic peak was obtained for the amorphous GaTeSn alloys, corresponding to the simultaneous crystallization of Te and Ga_2Te_5, they concluded that the addition of Sn to the GaTe system modifies the crystallization process, superposes the crystallization of Te and Ga_2Te_5 phases and stabilizes the Ga_2Te_5 phase in a wider range of temperatures, from 420 K. Moreover, the addition of small quantities of Sn to the $Ga_{20}Te_{80}$ binary alloy generates more stable amorphous phases. The activation energy, crystallization temperature, and crystallization enthalpy are also reported.

The aim of this work is to analyze the structure of cast and amorphous alloys of the GaTeSn system by means of XRD and ^{119}Sn Mössbauer spectrometry.

2. Experimental

The as-cast alloys were prepared by the joint melting of the 4 N pure elements, in the exact stoichiometric compositions, in an argon atmosphere. The sample compositions are given in Table 1. The homogeneity of the samples was determined by metallographic examination. All the samples were rapidly quenched with the piston and anvil technique [12] from the melt at temperatures between 600 and 800 °C. Samples in near circular form with a 30 μm average thickness were obtained. All the samples were analyzed by XRD using monochromatized Cu (Kα) radiation in a Rygaku θ-θ diffractometer. The ternary samples were analyzed by Mössbauer spectrometry, at 77 K, using a $Ca^{119}SnO_3$ source with transmission geometry. In all the samples, the isomer shift is reported relative to $CaSnO_3$ at 300K.

3. Results and discussion
3.1 As-cast samples

By means of XRD, equilibrium phases Te and Ga_2Te_3 were identified in samples A, B and C while Te and Ga_2Te_5 were detected in sample D. In this latter, a sizeable shift of the peak positions towards smaller angles corresponds to an increase of about 1-2 % in the lattice parameters. As shown in Table 2, the lattice parameters are larger than those of the pure compounds. Ga_2Te_5 is only detected in the as-cast D sample with larger amount of Sn (5 %). So, we conclude that the addition of this proportion of Sn stabilizes this phase up to room temperature. It is also important to emphasize that the SnTe phase was not detected by XRD in this sample, therefore we expect to find Sn in the compounds detected: Te and Ga_2Te_5.

Mössbauer spectra of cast samples exhibit single lines, as shown in Figure 1 while the refined values of the hyperfine parameters are listed in Table 3. These values correspond to divalent Sn in a

Table 1: Compositions of the studied samples (in at. %).

	Te	Ga	Sn
A	80.0	20.0	0
B	78.4	19.6	2.0
C	85.5	13.0	1.5
D	81.7	13.3	5.0

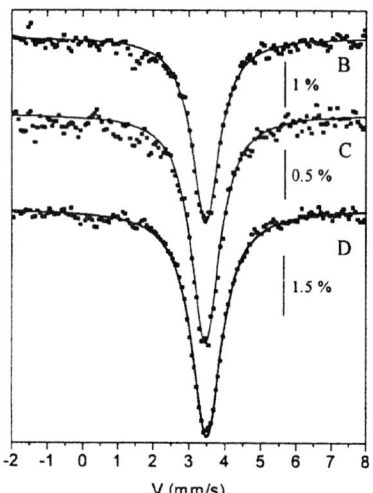

Figure 1: ^{119}Sn Mössbauer spectra of as-cast samples recorded at 77K.

Compounds and structural type	Sample B	Sample C	Sample D	crystalline phase
Te **(hexagonal)**	a = 4.47 ± 0.01 c = 5.95 ± 0.01	a = 4.47 ± 0.01 c = 5.98 ± 0.01	a = 4.51 ± 0.01 c = 6.02 ± 0.02	a = 4.457 c = 5.929
Ga_2Te_3 **(cubic)**	a = 5.934 ± 0.002	a = 5.937 ± 0.004	-	a = 5.898
Ga_2Te_5 **(tetragonal)**	-	-	a = 8.05 ± 0.01 c = 6.97 ± 0.01	a = 7.913 c = 6.848

Table 2: Lattice parameters (in Å) for the phases of the ternary samples. They are compared with those of the pure compounds.

cubic surrounding and are similar to those obtained for both phases SnTe and $Ga_2Te_3(^{119}Sn)$.

Nasredinov et al [5] proposed that the $Ga_2Te_3(Sn)$ lattice allows two different positions with cubic symmetry for Sn: a) octahedral interstitials and b) tetrahedral cation vacancies. At the octahedral interstitial sites, a Sn atom should collect stoichiometric cation vacancies in its second coordination sphere and form a regular octahedron of Te atoms in its first coordination sphere with the SnTe distance of 2.94 Å. The Sn atom occupying a cation vacancy should possess four neighboring Te atoms forming a regular tetrahedron.

In the SnTe compound, the surrounding of Sn atoms is a regular octahedron of Te atoms with the SnTe distance of 3.15 Å. Nasredinov et al [5] associate the Mössbauer parameters measured in $Ga_2Te_3(Sn)$ to the results for SnTe and proposed that the Sn atom is in the first of the possible sites: octahedral interstitials with an isomer shift around 3.55 mm/s. We obtain an identical conclusion for as cast samples B and C, where the same phase was detected.

One notes that the surroundings of Sn atoms in the measured samples do not depend on the crystalline phases detected by XRD. The crystalline phases identified in samples B and C are different from those of sample D. However, the Mössbauer parameters are similar in the three samples. So, one concludes that divalent Sn atoms are centered in a Te octahedral unit in these 3 compounds. Then, the structure of Ga_2Te_5 reveals a position of Sn with such a surrounding, consistent with a quasi-octahedral interstitial site: Sn has four neighboring co-planar Te atoms at 3.03 Å and two Te atoms at 3.42 Å forming a distorted octahedron.

Table 3: Mössbauer parameters for cast samples. IS and Γ are the isomer shift and line width, respectively.

sample	IS (mm/s) ± 0.02	Γ (mm/s) ± 0.02
B	3.47	0.93
C	3.46	0.95
D	3.49	1.02

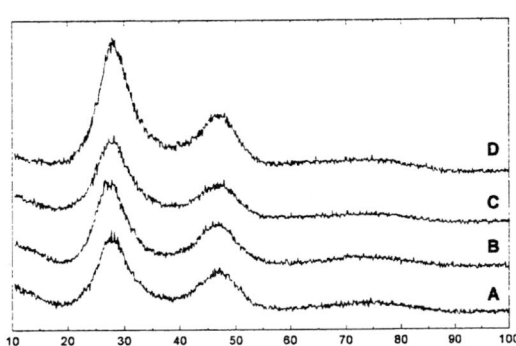

Figure 2: X-ray (Cu Kα) patterns of the rapidly quenched samples.

sample	Amorphous Phase					
	1st halo		2nd halo		3rd halo	
	$Q(Å^{-1})$ ±0.05	FWHM(°) ±0.5	$Q(Å^{-1})$ ±0.03	FWHM(°) ±0.5	$Q(Å^{-1})$ ±0.03	FWHM(°) ±0.5
A	1.96	5.5	3.26	8.0	4.90	9.5
B	1.95	5.7	3.22	6.6	4.87	10
C	1.97	6.4	3.27	6.8	4.94	8.5
D	2.00	5.6	3.24	6.3	4.94	8.3

Table 4: Structural parameters of amorphous phases: Q (in $Å^{-1}$), with $Q = (4\pi/\lambda) \sin(\theta)$, and the full width of half maximum FWHM (in °) of the 1st, 2nd and 3rd peaks of the X-ray patterns are reported.

3.2 Rapidly quenched samples

All the rapidly quenched samples are fully amorphous. The X-ray patterns which are shown in Figure 2, exhibit the presence of three broad peaks whatever the sample is, due to the amorphous phase. The X-ray characteristic patterns are reported in Table 4. The positions of the peaks are composition independent. The second peak at Q = 3.25 $Å^{-1}$, is found similar to that of the pure liquid Te [13].

The Mössbauer spectra of the rapidly quenched samples which are shown in Figure 3, display two broad, overlapping and asymmetrical lines, the intensities of which are composition dependent. The two lines which are approximately centered at 3.5 and 2.2 mm/s, are attributed to two components, consistent with two possible surroundings of Sn atoms. The spectra were fitted using two quadrupolar splitting distributions, with a linear dependence of IS on QS to take into consideration the asymmetry of the lines. The hyperfine parameters for each Sn site as well as the relative area of the spectra are reported in Table 5.

The first component is assigned to an octahedral Sn site because the IS values correspond to those obtained for the crystalline samples. The second component may be explained considering the work of Nasredinov et al [5]. They studied the Ga_2Te_3 crystalline phase doped with Sn, and a $Ge_{14.5}Sn_{0.5}Te_{85}$ amorphous sample. They reported that the Mössbauer spectra of this amorphous sample are single lines with an isomer shift of 2.07 mm/s at 80 K, and proposed that the surrounding of Sn atoms consists of a randomly distorted regular tetrahedron of Te atoms.

The addition of Sn to GaTe increases the glass forming range of the binary system. ^{119}Sn Mössbauer spectrometry allows to explain this result pointing out the existence of Te-Sn associations.

At this stage, we want to correlate the structure of the amorphous alloys to the results of the crystallization kinetics [11].

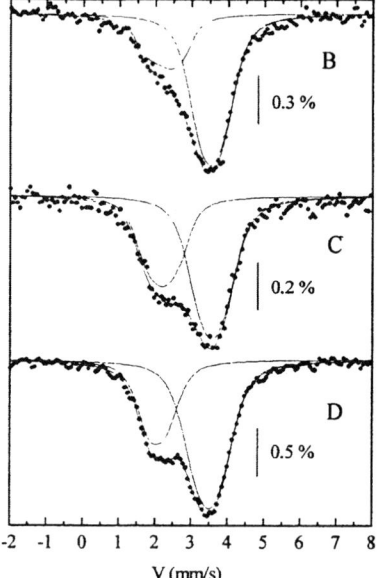

Figure 3: ^{119}Sn Mössbauer spectra of rapidly quenched samples recorded at 77K.

sample	First Sn site				Second Sn site				A_{12}
	<QS> (mm/s) ±0.01	σ(QS) (mm/s) ±0.01	<IS> (mm/s) ±0.01	σ(IS) (mm/s) ±0.01	<QS> (mm/s) ±0.01	σ(QS) (mm/s) ±0.01	<IS> (mm/s) ±0.01	σ(IS) (mm/s) ±0.01	
B	0.58	0.43	3.54	0.01	0.74	0.44	2.24	0.02	2.3
C	0.61	0.46	3.57	0.01	0.67	0.42	2.19	0.01	1.7
D	0.67	0.48	3.48	0.01	0.55	0.35	2.08	0.01	1.9

Table 5: Mössbauer parameters at 77K of rapidly quenched samples: average values of the quadrupolar splitting <QS> and isomer shift <IS> of the two Sn sites in the amorphous phases, both distribution widths, σ(QS) and σ(IS) and the area of the first distribution relative to the second one A_{12}.

a) Although the structure of amorphous alloys detected by XRD does not depend on the composition, one concludes that Sn atoms play a fundamental role in the glass forming ability, increasing both its composition and temperature ranges.

b) Comparing amorphous alloys A and B, containing the same at. % Ga, it is concluded that Sn addition to GaTe system increases the glass and crystallization temperatures in 70 K, as well as the apparent activation energy of the amorphous phases, and superposes the crystallization peaks detected by DSC [11]. Such a feature is consistent with the presence of cluster-like Ga_2Te_3 associations in the melt and the amorphous phase [16]. The analysis of the structure of compounds Ga_2Te_5 and Ga_2Te_3 leads us to model the structure of the amorphous phase. Ga_2Te_5 is formed by chains of $GaTe_4$ tetrahedra connected by Te atoms while Ga_2Te_3 is based on a tetrahedral atomic coordination as in the zinc blende crystal. Both structures have an interstitial site with an octahedral coordination for the Sn atom: in such a way, Sn helps to join the tetrahedron chains formed by Te and Ga and favors consequently the stability of the structure of these compounds. On the other hand, tetrahedron GeT_4 (T = S, Se) chains connected by tin atoms constitute the amorphous structure of other chalcogenide systems GeSnT [14, 15]. Consequently, we propose that chains of distorted tetrahedra form the binary amorphous Ga_xTe_{100-x} structure. Addition of Sn atoms stabilizes the binary amorphous structure because they generate new bonds between the chains. Then, if the atomic % of Ga is the same, the ternary amorphous GaTeSn structure is more stable.

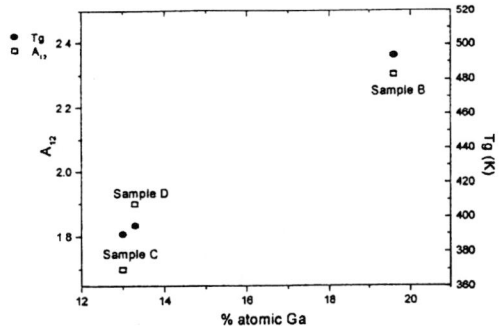

c) To analyze the results obtained on ternary samples B, C and D, we first consider that the thermal stability of ternary GaTeSn amorphous phase is

Figure 4: Glass temperatures Tg, obtained by DSC [11], and Mössbauer spectra area of the first site distribution relative to the second one A_{12} are plotted versus the atomic fraction Ga of the ternary samples.

connected with the atomic % of the major components (for example Te or Ga), that can be linked with the fraction of tetrahedral Sn sites. The glass transition temperature T_g reported in [11], and the Mössbauer spectra area of the first site distribution relative to the second one A_{12} are plotted versus the atomic fraction Ga in the ternary samples and are shown in Figure 4. This result may be compared with those of the molecular structure of amorphous samples Ge_{2-2x} Sn_{2x} Se_3 studied by DSC and Mössbauer spectrometry [14]. Enzweiler and Boolchand observed that T_g exhibits a maximum for x=0.5 and considered this fact may be a consequence of a pronounced polymerization or connectivity of the glass network.

On the other hand, considering that: 1) the Ga_2Te_3 associations are present in the amorphous phase and 2) the Sn atom in the crystalline structure Ga_2Te_3 can be located either in tetrahedral positions which are vacancies in the chains of tetrahedra, or octahedral sites connecting the chains, then, the smaller amount of tetrahedral sites in the amorphous phase could be attributed to the chains having less vacancies. Also, the amorphous structure should be more compact. In other words, the higher values of A_{12} could indicate that the coupling of the chains in the glass network is enhanced, increasing its stability. Sn controls the binary amorphous phase structure, even if it occurs as a minor species in the present alloys.

d) The Ga_2Te_5 compound is thermodynamically favored to Ga_2Te_3, both in as cast sample D and in the products of crystallization. Such a feature confirms that the addition of Sn, above a threshold concentration, stabilizes the chains of tetrahedral $GaTe_4$ units of the present systems.

4. Conclusions

In the crystalline samples the divalent Sn atoms are always surrounded by Te in octahedrally coordinated environments. It is observed that the addition of Sn, above a threshold concentration, stabilizes the high temperature phase Ga_2Te_5 at room temperature. This idea is verified considering that this phase is detected as a crystallization product of amorphous phases at lower temperatures. In the present amorphous samples, the Sn environments correspond to associations of Sn and Te with octahedral and tetrahedral coordination.

The small addition of Sn to the $Ga_{20}Te_{80}$ binary composition generates more stable amorphous phases because Sn atoms help to join the chains of tetrahedra formed by Te and Ga atoms and favor the stability of the ternary amorphous samples which increases with the rate of octahedral Sn sites relative to tetrahedral ones.

Although the presence of the tetrahedral sites, which are not observed in the crystal, evidences the necessity of an additional diffusive process to allow nucleation and growth of the equilibrium phases, it is concluded that the stability of the amorphous phase fundamentally depends on the linkage ability of Sn atoms in octahedral environments.

References

[1] S.K. Hsiung and R. Wang , J. Appl. Phys. 49 (1) (1978) 280.

[2] G. Parthasarathy, S. Asokan, S.S.K. Titus and R.R. Krishna, Phys. Letters A 131 (7,8) (1988) 441.

[3] H.L. Luo and P. Duwez, Appl. Phys. Lett. 2 (1963) 21.

[4] R. Wang, Bulletin of Alloy Phases Diagrams 2, 3 (1981) 269.

[5] F. S. Nasredinov, V. Masterov, C. Saidov, P. Seregin, N. Troitskaya and H. Tschirner, Phys. Stat. Sol.(a) 107 (1988) 291.

[6] E. Irle, B.Gather, R.Blachnik, U.Kattner, H.Lukas and G.Petzow, Z. Metallkde 78 8 (1987) 535.

[7] R. Sharma and Y. Chang, Bulletin of Alloy Phase Diagrams 7, 1 (1986) 72

[8] I. Hatta and A. Ikushima, Jpn. J. Appl. Phys. 20 (1981) 1995.

[9] O. Valassiades and N. Economou, Phys. Status Solidi 30 (1975) 187.

[10] V. Fano and I Ortalli, J. Chem. Phys. 61, 12 (1974) 5017.

[11] M.Fontana, B. Arcondo, M.T. Clavaguera-Mora and N. Clavaguera, Non-Crystalline and Nanoscale Materials, Proc. V International Workshop on Non-Crystalline Solids, Ed. J. Rivas and M.A. López-Quintela, Word Scientific, Singapore (1998), p. 343-348.

[12] G. Quintana, H. Sirkin, M. Rosen, D. Kurlat and E. Frank, Revista Brasileira de Fisica 9 (1979) 1.

[13] A. Mueller, W. Hoyer, E. Thomas and M.Wobst, Phys. Stat. Sol. (a) 84, K97 (1984).

[14] R.N.Enzweiler and P.Boolchand, Hyper. Interac. 27 (1986) 393.

[15] P.Boolchand, Hyper. Interac. 27 (1986) 1.

[16] Y.Waseda, The Structure of Non-Crystalline Materials (Mc Graw Hill Inc., USA, 1980) 168.

EFFECT OF SOME PLANT EXTRACTS ON
THE ATMOSPHERIC CORROSION OF CARBON STEEL

C. Hernández[1], E. García de Saldaña[1] and J. A. Jaén[2].
[1]Maestría en Ciencias Químicas, Universidad de Panamá.
*[2]Centro de Investigaciones con Técnicas Nucleares / Depto. de Química, Universidad de
Panamá.*

The rust formed on a mild-carbon-steel (A-36) exposed in an urban environment have been
investigated using Mössbauer spectroscopy. Clean and rusted steel panels were treated with
15% w/V solutions of the aqueous extracts from the plants: *Opuntia elatior* mill.,
Acanthocereus pentagonus (L.) Britton, *Mimosa tenuiflora*, *Bumbacopsis quinata* (Jacq.)
Dugand and *Acacia mangium* Willd. The treated steels showed a significant slowing down of
atmospheric corrosion rates attributed to the conversion reaction of steel and rust with the plant
extracts solutions.

The dominant phases present in all samples consisted of non-stoichiometric magnetite
($Fe_{3-x}O_4$), goethite (α-FeOOH) of intermediate particle size, lepidocrocite (γ-FeOOH) and
superparamagnetic particles.

1. INTRODUCTION

A number of tannins and related polyphenols have been found suitable for improving
protection of steel and rusted steel [1-7] against further corrosion. This characteristic can
be used to develop atmospheric anticorrosive coatings based on the formation of iron
tannates. A tannin-based solution may react directly with a rusted surface to form a water-
insoluble, reticulated protective layer of a tannin-iron complex that can be topcoated.
Several oxide conversion formulations based on tannins have been shown to provide good
corrosion protection in different climates [2,3,8-12]. On the other hand, the performance of
tannic acid solutions for protection of rusted steel prior to painting has been question due to
the high solubility of the films and low anti-corrosion efficiency [13,14]

Identification of corrosion products, the rate of atmospheric corrosion, and the effect of
different plant extracts on the atmospheric corrosion of clean and rusted steel are of
interest. We have recently started a study on the feasibility of application of water extracts
of some Panamanian plants as atmospheric corrosion inhibitors.

Mössbauer measurements in transmission geometry were used to follow the phase
variations. We also report on the corresponding corrosion rates.

2. EXPERIMENTAL

15% (w/V) solutions of dry powders of the aqueous extract of the plants *Opuntia elatior*
mill., *Acanthocereus pentagonus* (L.) Britton, *Mimosa tenuiflora*, *Bumbacopsis quinata*
(Jacq.) Dugand and *Acacia mangium* Willd. were evaluated as atmospheric corrosion
inhibitors of mild-carbon-steel (A-36). The extracts were obtained as described separately.
Commercial tannic acid (Merck) was also used for comparison purposes. After cleaning
and degreasing with acetone, steel panels were exposed for a 12-month period in a station
located on the University Campus in Panama City, 2 km from the seashore. The site is
considered to be an urban environment classified according to ISO 9223 as C3 [16].

Four type of samples were used for phase analysis. First, rust samples were removed

Table 1. Program of exposure of panels for phase analysis

	3 month	6 months
December 1996 - February 1997	X-3-1 CO-3-1	X-6-1 X-6-2
March 1997 - May 1997	X-3-2 CO-3-2	CO-6

X = plant extracts solutions

Opuntia elatior mill.	OE
Acanthocereus pentagonus (L.) Britton	AP
Mimosa tenuiflora	MT
Bumbacopsis quinata (Jacq.) Dugand	BQ
Acacia mangium Willd.	AM
Tannic acid	TA

from control panels denoted as CO-3-1, CO-3-2 and CO-6 in Table 1, after exposure to the atmosphere. All other panels were given an indoor brush application of 2 ml of the 15% (w/V) solutions of the plant extracts on each side of the 150 x 100 x 3 mm steel panels, be allowed to dry for 48 hours, and subsequently expose to the atmosphere. The second type of samples (X-3-1 and X-3-2) were exposed out-of doors for three months, whereas the third (X-6-1) and fourth type (X-6-2) were exposed for six months. Rusted steels of the latter type were repainted outdoor after three months of exposure.   The program is summarized in Table 1.

After exposure, corrosion rates were determined as corrosion penetration (p) from weight losses using ISO 9226 [17]. A conventional constant-acceleration Mössbauer spectrometer was used for recording the spectra at room temperature. The evaluations were carried out by least-squares fitting procedure and by obtaining the hyperfine-field distribution with the program NORMOS.

3. RESULTS AND DISCUSSION

The weight loss values, expressed as penetration, obtained from samples exposed for up to a year, were plotted vs. time, in bilogarithmic form, obtaining the experimental equation ln p = 3.898 + 0.405 ln t, with a correlation coefficient of 0.973. The slope is within expected values for carbon-steels, as could be observed in a previous work [16] for a mild-carbon-steel (SAE-1020) tested in a similar station but for longer periods. It is related to the influence of the environment on corrosion rate. The intercept has the physical meaning of the thickness loss after the first year, being very high in this case.

In order to determine the corrosion resistance effectiveness of the different plant extracts, non-treated and painted specimens were exposed to the atmosphere as described in the experimental part.   The results given in Figure 1 indicate that the overall performance of the plant extracts is better than that of control panels. The corrosion protection of the plant extracts were found in the order TA=AM > BQ = AP > MT > OE. It is interesting to note that, in general, exposed rusted panels, painted again and then re-exposed perform better. This is an indication of the convenience of coating rusted surfaces to obtain better protection.

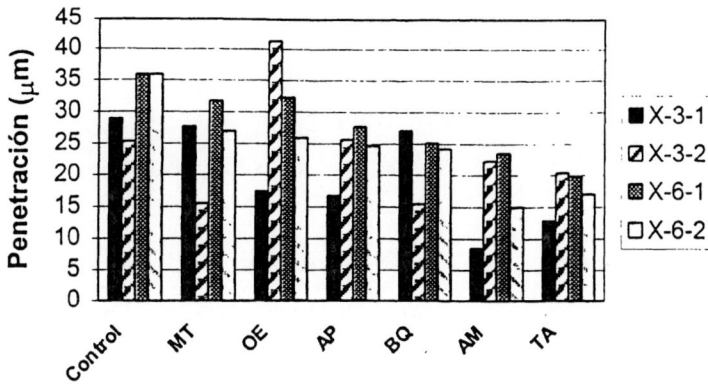

Figure 1. Weight loss values, expressed as penetration, of the non-treated and painted steel specimens exposed to the atmosphere.

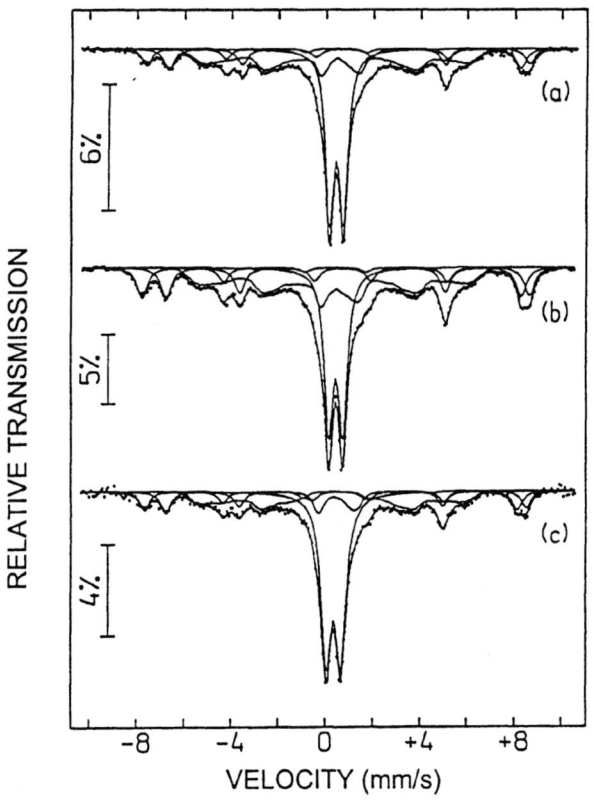

Figure 2. Room temperature Mössbauer spectra of rust from samples (a) CO-6, (b) TA-6-1 and (c) TA-6-2

Table 2. Room temperature Mössbauer parameters and relative areas (%)

Spectral Component	G*	M_1	M_2	Q
Hyperfine magnetic field H / kOe	-	498-507	459-466	-
Quadrupole splitting Δ or E_q / mm s^{-1}	-	-0.01-0.22	0.00-0.04	0.58-0.61
Isomer shift δ / mm s^{-1}	-	0.31-0.40	0.64-0.68	0.33-0.39
Line width Γ / mm s^{-1}	-	0.44-0.72	0.43-0.77	0.42-0.52
C0-3-0-1	26.2	14.0	17.2	42.6
C0-3-0-2	28.0	10.5	16.0	45.5
C0-6-0	36.0	8.8	14.2	41.0
A1-3-1-1	28.8	11.9	15.0	44.3
A2-3-1-1	27.0	16.4	14.6	42.0
A3-3-1-1	29.9	15.7	14.0	40.4
A4-3-1-1	25.1	13.0	18.0	43.9
A5-3-1-1	26.9	13.4	16.3	43.4
A6-3-1-1	24.7	14.4	19.1	41.8
A1-3-1-2	25.5	9.8	18.6	46.1
A2-3-1-2	29.9	18.2	12.4	39.5
A3-3-1-2	28.1	7.8	13.0	51.1
A4-3-1-2	40.8	12.5	18.7	28.0
A5-3-1-2	24.4	18.2	21.0	36.4
A6-3-1-2	27.5	17.8	16.5	38.2
A1-6-1	45.4	7.4	9.8	37.4
A2-6-1	39.7	10.4	10.5	39.4
A3-6-1	38.7	12.2	11.0	38.1
A4-6-1	38.1	9.2	11.8	40.9
A5-6-1	41.9	8.6	14.0	35.5
A6-6-1	38.1	11.7	17.1	33.1
A1-6-2	30.8	7.5	11.1	50.6
A2-6-2	29.8	11.0	16.2	43.0
A3-6-2	32.1	12.2	13.6	42.1
A4-6-2	38.4	11.7	12.2	37.7
A5-6-2	42.2	8.2	11.7	37.9
A6-6-2	28.2	10.0	13.3	48.5

*Fitted as a simple distribution of hyperfine fields

G = goethite, Q = quadrupole doublet and M = hyperfine magnetic splitting

The Mössbauer spectra of the corrosion products removed from all samples consisted of a quadrupole doublet, the characteristic two superimposed sextets of the spinel phase non-stoichiometric magnetite ($Fe_{3-x}O_4$) and a broad not well resolved magnetic pattern. Representative room-temperature Mössbauer spectra are shown in Figure 2. The quadrupole doublet (indicated as Q in Table 2) could be associate to a mixture of lepidocrocite γ-FeOOH with superparamagnetic particles [16].

The broad not well resolved magnetic pattern was attributed to goethite α-FeOOH of intermediate particle size (20-100 nm) [18]. The size distribution of goethite is responsible for the lineshape. This is related to a hyperfine-field distribution, thus the spectral contribution was fitted using a simple distribution.

We have previously shown [16] that intermediate particle size goethite contributes significantly to formation of protective layers, up to a limit given by the saturation of the surface coverage. Assuming validity of the already used Langmuir-type approach, a plot of p/A_G vs. p, where A_G is the room temperature Mössbauer area of goethite of intermediate particle size, reveals two different type of behavior, as shown in Figure 3. A straight line with higher slope fits better all data in which the steel samples, clean or rusted, had been coated with the plant extracts solutions three months before. This might be related to a diminished goethite content, which in turn may be associated to the conversion reactions of steel and rust with the plant extracts solutions to give mono- and bis-type complexes [15]. In both cases, the complexation reaction with the steel substrate (Fe) or with the outermost rust surface (lepidocrocite), iron-plant extracts insoluble complexes similar to ferric tannates are obtained. Such films contribute to the further corrosion retardation, and at the same time interfers with the formation of goethite. It is interesting to note that γ-FeOOH is the most reactive rust component, giving ferric tannates in the reaction with aqueous solutions of tannic acid and related compounds [19,20]. The room temperature Mössbauer parameters of the tannates are similar to the doublet attributed to lepidocrocite and superparamagnetic particles [15,21].

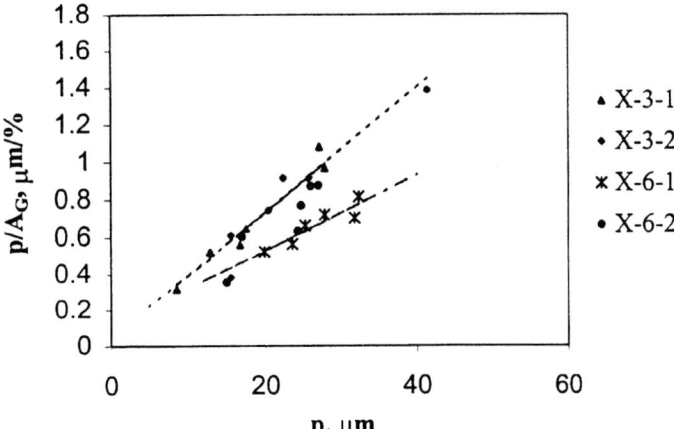

Figure 3. Langmuir-type of fitting of spectral-area of intermediate particle size goethite.

4. CONCLUSION

The reaction of aqueous plant extracts with clean and rusted steel provides some protection to atmospheric corrosion.

An insoluble tannate-like film is formed on the surface. Such films contribute, along with goethite, to the surface coverage and corrosion retardation.

Rust consisted mainly of non-stoichiometric magnetite ($Fe_{3-x}O_4$), goethite (α-FeOOH) of intermediate particle size, lepidocrocite (γ-FeOOH) and superparamagnetic particles.

ACKNOWLEDGEMENT

The work was supported by the program UNIPAN-BID IV (University of Panama-International Development Bank).

References

[1] E. Knowles and T. White, J: Oil Col. Chem. Assoc. 41 (1958) 10.
[2] A.J. Seavell, J.Oil Col. Chem. Assoc. 61 (1978) 439.
[3] T.K. Ross and R.A. Francis, Corros. Sci. 18 (1978) 351.
[4] D. Vacchini, Anti-Corrosion, 32 (1985) 9.
[5] P.J. DesLauriers, MP 26 (1987) 35.
[6] A.J. Seavell, J. Oil Col. Chem. Assoc. 75 (1992) 293.
[7] G. Matamala, W. Smeltzer and G. Droguett, Corrosion 50 (1994) 270.
[8] M.I. González and R. López Planes, Rev. Iberoam. Corros. Prot. 19 (1988) 374.
[9] R. Franiau, Congr. Colloq. Univ. Liege, 74 (1976) 129.
[10] E.A. Kanevoskaya and A.I. Katova, Lakohrasoch. Mater. Ikh Primen 5 (1970) 31.
[11] G. Wieczorek and C. Krepski, Ochr. Koroz 19 (1976) 98.
[12] L. Sorinas, F. Luzardo, T. Ochoa, E. Caraballo, A. Cabezas, L. Vargas and M. García, Corros. Prot. Mater 16 (1997) 19.
[13] M. Morcillo, S. Feliu, J. Simancas, J.M. Bastidas, J.C. Galván, S. Feliu Jr. and E.M. Almeida, Corrosion 48 (1992) 1032.
[14] J.C. Galván, S. Feliu Jr., J. Simancas, M. Morcillo, J.M. Bastidas, E. Almeida and S. Feliu, Electrochim Acta 37 (1992) 1983
[15] J. A. Jaén, E. García de Saldaña and C. Hernández, submitted to Hyp. Int.
[16] J. A. Jaén, M. Sánchez de Villalaz, L. de Araque and A. de Bósquez, Hyp. Int. 110 (1997) 93.
[17] ISO 9226, *Corrosion of metals and alloys. Method for determination of corrosion rate of standard specimens for the evaluation of corrosivity* (International Standards Organization, Geneva, 1991).
[18] C. Barrero, R. Vandenberghe, E. De Grave and G. Pérez, Rev. Col. Fis. 27 (1995) 387.
[19] J. Gust, Corrosion 47 (1991) 453.
[20] J. Gust and I. Wawer, Corrosion 51 (1995) 37.
[21] J. Gust and J. Suwalski, Corrosion 50 (1994) 355.

STUDY OF THE INITIAL CORROSION PRODUCTS FORMED ON CARBON STEEL EXPOSED ALONG GULF OF MÉXICO

Rama Balasubramanian[1], D. C. Cook[1], T. Perez[2] and J. Reyes[2]

[1]Department of Physics, Old Dominion University, Norfolk, VA 23529, U. S. A.

[2]Programa de Corrosión del Golfo de México, Universidad Autónoma de Campeche, Campeche, México.

High mean annual temperature, humidity, time-of-wetness and very high atmospheric pollutants are some of the critical factors that affect the performance of steels. Several regions along the Gulf of México have highly corrosive environments that are detrimental to the steel structures in those areas. A systematic study has been made of the initial corrosion products which form on carbon steel exposed in marine and near marine locations, in México. Two sets of coupons were exposed up to 12 months at Programa de Corrosión del Golfo de México and Servicio Meteorológico Nacional located 300 m and 4000 m respectively from the shoreline at Campeche, México. The resulting corrosion products formed on the coupons were characterized as a function of exposure time and environment by Mössbauer spectroscopy and x-ray diffraction. X-ray diffraction analysis showed that the corrosion products predominantly contained lepidocrocite and goethite as the two main oxide constituents. Transmission Mössbauer analysis at room temperature showed the presence of only a paramagnetic doublet, characteristic of fine particle goethite and lepidocrocite. The superparamagnetic doublet transformed into a broadened sextet of goethite at 77 K, indicating a distribution of particle size in the range of 8-15 nanometers. As a function of exposure time, the average hyperfine field of goethite measured at 77 K showed and increase. This indicated that the particle size increased as a function of exposure time. The decrease in the distribution of the hyperfine field of goethite measured at 77 K indicated that the particle size became more uniform as the exposure time increased.

1. INTRODUCTION

Steel has been the most widely used alloy for structural and industrial applications, since the beginning of the industrial revolution. The lifetime of steel is directly related to its performance against corrosion. Corrosion of steel caused by environmental conditions is termed as atmospheric corrosion. Little is known about the effect of atmospheric corrosion on the performance of steel due to the many environmental parameters which control the corrosion. Systematic studies on the formation and transformation of corrosion products and their correlation to the environmental parameters is very important to understand the processes that lead to the atmospheric corrosion of steels. Such research would lead to improving the performance of steels in highly corrosive environments.

Many regions along the Gulf of México have extremely corrosive environments due to high mean annual temperature, humidity, time-of-wetness and very high atmospheric pollutants. Since carbon steel is most widely used for structural purposes in México, we have addressed the effect of short-term atmospheric corrosion on carbon steel exposed along the Gulf of México. The process of formation and the possible transformation of corrosion products resulting from short-term exposure of carbon steel, both as a function of environmental conditions and exposure time, has been investigated. This paper presents the results of the study of atmospheric corrosion of carbon steel, exposed for less than 12 months at two different locations in Campeche, México.

2. SAMPLE PREPARATION AND EXPERIMENTAL PROCEDURE

The carbon steel coupons measuring 150 x 100 x 6 mm were subjected to atmospheric exposure at sites Programa de Corrosión del Golfo de México (PCGM) and Servicio Meteorológico Nacional (SMN), in Campeche , México. The exposure sites PCGM and SMN located 300 m and 4000 m inland respectively from the coast of Gulf of México, were assigned site identification tags of CP3 and CP2 . The coupons were subjected to atmospheric exposure for less than 12 months. Additional sets of coupons were exposed at these two locations at the beginning of the different seasons as well, to monitor the effect of seasonal variation on the initial atmospheric corrosion. Table 1 presents the critical environmental factors namely, temperature, humidity and chloride concentrations measured during the exposure period. Following the exposures, the surface of all the coupons, except CP3-08, appeared to be completely covered by a thick layer of corrosion products with no steel substrate visible. The corrosion layer on coupon CP3-08 was thin. In-situ x-ray diffraction analysis was performed to identify the oxide phases present, using a Phillips x-ray diffractometer with Cu K_α radiation of wavelength of 1.54056 Å. Transmission Mössbauer Spectroscopy (TMS) was used to record the spectra of hyperfine interactions, characteristic of the oxides, at room temperature (300 K), liquid nitrogen temperature (77 K) and liquid helium temperature (4 K). For the TMS analysis, the corrosion coatings were removed from a 50 x 25 mm piece of the exposed coupons, by carefully scraping the rust layer off the steel substrate. About 15 mg of the coating was mixed with 150 mg of boron nitride binder, and pressed into a 1 cm diameter tablet. The Mössbauer spectra were recorded using a 50 mCi Co/Rh source.

Table 1. Environmental conditions at the two exposure locations.

Sample	Location	Exposure Time (Months)	Exposure Period	Average Temp. (°C)	Humidity (%)	Chloride Conc. (mg/m² day)
CP3-04	PCGM	2	May 97-June 97	29.7	67	5.12
CP3-08	PCGM	3	May 97-July 97	29.2	70	13.60
CP3-14	PCGM	6	May 97-Oct 97	28.6	74.5	12.50
CP3-18	PCGM	9	May 97-Jan 98	27.3	75.7	12.00
CP2-32	SMN	6	May 97-Oct 97	28.6	74.5	13.17
CP2-36	SMN	9	May 97-Jan 98	27.3	75.7	17.23

3.   RESULTS AND DISCUSSION

In-situ x-ray diffraction patterns for all the coupons were recorded in the angular range of 10-120 deg to obtain a basic identification of the oxides present in the corrosion coating. Figure 1 shows the x-ray diffraction patterns for the four corroded coupons in the angular range of 45-65 degrees. A shift of ±0.09 degrees was observed in the diffraction patterns of the corroded coupons, as compared to α-Fe peaks used for calibration. This was attributed to sample displacement error, since the coatings analyzed were still attached to the steel. Comparison of the x-ray patterns of the coatings with those of the standard oxides, indicated that lepidocrocite, and goethite were the dominant oxides present in the coatings [1]. X-ray diffraction patterns corresponding to the steel substrate were also observed. The mean peak width of the x-ray patterns of the coupons showed an increase, as compared to the pure standard oxides. This clearly suggested that the particle size was much smaller than the standards whose average particle size was 100 nm. The absence of the diffraction peak at 56.36 degrees, characteristic of the akaganeite phase was very surprising indicating that akaganeite did not form in the marine environment following short-term exposure.

R. Balusubramanian et al. / Study of the initial corrosion products formed on carbon steel exposed along Gulf of Mexico

39

Transmission Mössbauer analysis at room temperature of the corroded coupons was characterized by only a doublet. It was not possible to separate the contribution of the constituent oxides to the doublet at room temperature. The room temperature spectra were fitted to a distribution program with two doublets to improve the quality of the fits with reduced χ^2 values. The two doublets differ noticeably in their quadrupole splittings (0.690 ± 0.014 and 0.56 ±0.018 mm/s) and distribution widths (0.24 ±0.02 and 0.060 ±0.004 mm/s). However, the isomer shifts for the two doublets (0.37 and 0.36 mm/s) do not vary significantly. The fit parameters suggested that the corrosion products could contain lepidocrocite, akaganeite and/or superparamagnetic goethite [2] . But the x-ray analysis indicated the absence of akaganeite in the corrosion products. Figure 2 shows the TMS spectra at 77 K of the corrosion products. The low temperature spectra consisted of two components namely, a doublet and a broadened sextet. The doublet was fitted to a distribution and the fit parameters yielded an average quadrupole splitting of $\Delta = 2\varepsilon = 0.620±0.025$ mm/s, with a distribution of $\sigma_{(\Delta)} = 0.20±0.06$ mm/s .The fit parameters indicated the presence of lepidocrocite and /or fine particle goethite. The final fit parameters for the broadened sextet corresponded to that of magnetically ordered goethite having a distribution of particle size. For the coupon exposed for 3 months at PCGM, the average mean field was determined to be 386.8 kOe. Table 2 presents the Mössbauer fit parameters of the spectra recorded at 77 K. The low temperature Mössbauer results, clearly showed that the goethite displayed a distribution of magnetic field, which then indicated a distribution of particle size categorized as P1 and P2, where P1 represents particle size in the range 8-15 nm and P2 represents particle size < 8 nm . Since the magnetic order for the goethite was not present at room temperature, the particle size was proposed to be in the range of 8-15 nm and corresponded to P1 [2]. Also the distribution of the doublet measured at 77 K strongly indicated the presence of ultra-fine particle goethite

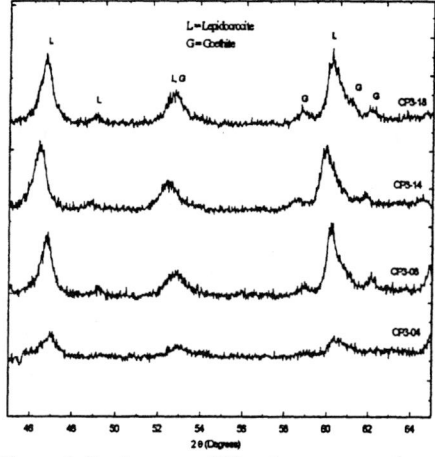

Figure 1. In-situ x-ray diffraction patterns of the corroded coupons exposed at PCGM.

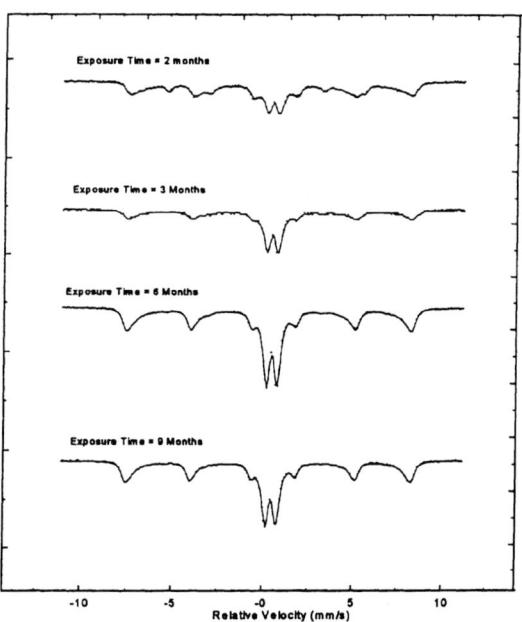

Figure 2. Transmission Mössbauer spectra recorded at 77 K for the carbon steel coupons exposed at PGCM.

categorized as P2 with the particle size <
8 nm [3]. Goethite with particle size in
this range was still superparamagnetic at
77 K. The relative contribution of doublet
increased as a function of exposure time,
while that of the magnetically ordered
goethite decreased. TMS analysis at 4K
was performed on the sample CP3-18.
The resulting spectrum was fitted with a
distribution program. The fit parameters
showed a the presence of three
magnetically split sites. The hyperfine field
value of the first site was 446.9 kOe with
an isomer shift of 0.49 mm/s and
quadrupole splitting $\Delta = 2\varepsilon = 0.02$ mm/s
characteristic of small particle
lepidocrocite. The reduction in hyperfine
field of lepidocrocite recorded at 4 k was
a result of the particle size being much
smaller than 100 nm. The fit parameters
pertaining to the second site showed a
distribution of hyperfine field

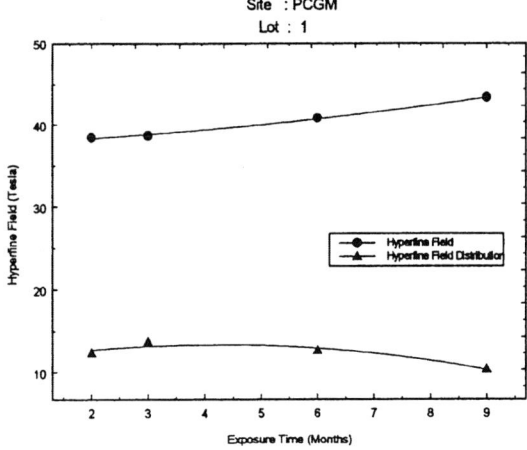

Figure 3. Variation of the mean hyperfine field and the distribution of the mean field of goethite at 77 K as a function of exposure time.

corresponding to goethite, with an average hyperfine field of 500.2 kOe. Of interest was the third site whose hyperfine field was 456.1 kOe, with an enormous distribution of 43.2 kOe. This phase was not identified by the x-ray analysis. But several earlier reports on ferrihydrite indicate that x-ray diffraction analysis of very poorly ordered material showed only the two *hk* lines at 2.5 and 1.5 Å that were broad[3]. Also the hyperfine field value of the two line ferrihydrite has been reported to be 465 kOe [4]. Unpublished data on micro-crystalline oxides by Cook etal., showed the presence of a similar phase, that was believed to be a precursor of the final oxides formed in the corrosion products. In this study, we believe that this x-ray opaque site observed at 4 K might be the precursor to the goethite phase, which could be ferrihydrite and /or ultra fine particle goethite with particle size <8 nm. The relative contribution of goethite, lepidocrocite to the 4 k spectra were 38.08 % and 11.45 % respectively, while that of the x-ray amorphous phase was 50.44 %. The ratio of α-FeOOH/γ-FeOOH was nearly 3.5. Figure 3 shows the relationship between the exposure time, and average field of the magnetically ordered goethite measured at 77 K for the coupons exposed at PCGM. As a function of exposure time the mean field of the magnetically ordered goethite increased. Also the distribution of hyperfine field decreased as a function of time. Since the mean field is indicative of the average particle size, it was concluded that as exposure time increased, the particle size increased. The reduction in the field distribution strongly suggested that, as the time of exposure increased, the particle size became more uniform. The results were consistent for the coupons exposed at different times of the year and different locations as well.

4.   CONCLUSIONS

The corrosion products formed on carbon steel coupons following short term atmospheric exposure at two locations along the Gulf of México were studied. In-situ x-ray diffraction and transmission Mosssbauer analysis of the coupons showed that lepidocrocite and goethite were the dominant oxides present in the corrosion products. Akaganeite was absent in the corrosion coatings. Transmission Mössbauer analysis at room temperature suggested the presence lepidocrocite and super-paramagnetic goethite. In addition, transmission Mössbauer analysis showed that, the super-para magnetic component became magnetically ordered at 77 K, which then indicated the particle size of goethite to be in the range of 8-15 nm. The doublet measured at 77 K comprised of lepidocrocite and ultra fine particle goethite whose particle size was less than 8 nm. As a function of exposure time, the average hyperfine field of goethite measured at 77 K showed an increase. This indicated that the particle size increased as a function of exposure time . The decrease in distribution of hyperfine field of goethite measured at 77 K indicated that the particle size became more uniform as the exposure time increased.

Table 2. Transmission Mössbauer fit parameters recorded at 77 K for magnetic goethite present in the carbon steel coupons.

Sample	IS(mm/s)	Δ (2 ε) (mm/s)	B (kOe)	σ (kOe)	Sextet Area(%)	Doublet Area(%)
CP3-04	0.473	-0.168	384.3	124.0	79.06	20.89
CP3-08	0.476	-0.190	386.8	137.5	63.68	36.39
CP3-14	0.473	-0.201	408.1	126.6	64.69	35.30
CP3-18	0.476	-0.214	433.9	103.7	62.10	37.89
CP2-32	0.473	-0.204	438.5	89.4	53.58	46.41
CP2-36	0.475	-0.218	430.3	103.7	51.07	48.92
Bulk Goethite [6]	0.475	-0.240	496.0	0	100	-

5.   REFERENCES

[1] D. C. Cook, S. J. Oh, and H. E. Townsend, The protective layer formed on steels after long-term atmospheric exposure, Corrosion 98, paper no. 343, (1998).
[2] L. H. Bowen, E. de Grave and R. E. Vandenberghe, in Mössbauer Spectroscopy Applied to Magnetism and Materials Science, vol. 1, ed G. J. Long (Plenum Press, New York, 1993) p. 115.
[3] E. Murad and J. H. Johnston, in Mössbauer Spectroscopy Applied to inorganic Chemistry, vol. 2, ed G. J. Long (Plenum Press, New York, 1978) p. 507.
[4] E. Murad, Magnetic properties of microcrystalline iron (III) oxides and related materials as reflected in their Mössbauer spectra, Phys. Chem. Minerals, vol 23, (1996) p. 248-262.
[5] E. de Grave, R. M. Persoons, D. G. Chambaere, R. E. Vandenberghe and L. H. Bowen, An [57] Fe Mössbauer Effect study of Poor Crystalline γ-FeOOH, Phys. Chem. Minerals, vol 12, (1986) p. 61-67.
[6] S. J. Oh, Characterization of iron oxides and atmospheric corrosion of steel, Old Dominion University, Ph.D Thesis, (1997) p.157-180.

Cu-DOPED MAGNETITE OBTAINED BY HYDROLYSIS

A. L. Morales[1,2], H. Mosquera[1] and C. Arroyave[2]

[1]*Departamento de Física, Universidad de Antioquia,*
A.A. 1226, Medellín, Colombia
[2]*Grupo de Corrosión y Protección, Departamento de Ingeniería de Materiales,*
Universidad de Antioquia, A.A.1226, Medellín, Colombia

Abstract

Weathering steels have been found to present a good corrosion resistance due to the presence of alloying elements like Cu and Cr. Under corrosion conditions they form a protective layer compose of different iron oxides including magnetite. To study the role of Cu we have produce 1% and 5% Cu-doped magnetite by hydrolysis of different iron salts and studied them by Mössbauer spectroscopy and X-ray diffraction. The largest increase in linewidth, with increasing Cu content, occurs at B-sites suggesting that Cu prefers to substitute Fe at that site. As a result of the Cu presence goethite also forms and its amount increases as the doping increases.

1. Introduction

Understanding the mechanism by which weathering steels create a protective layer is an important subject in corrosion science[1-6] and especially the role played by alloying elements. These elements may be incorporated in the oxides forming the rust, in amounts ranging from 5-10%, as found by Yamashita et al.[4] and Jaen et al.[7]. Davalos et al.[1] have found that the protective layer formed in these steels is compose mainly of superparamagnetic goethite, Yamashita et al.[4] found a composition of Cr-substituted goethite or alternatively alternating layers of lepidocrocite and goethite and Oh et al.[6] found that it is composed of goethite and hematite. Nasradazani and Raman[3] have found that magnetite is the main rust component under continuous wet conditions and Oh et al.[6] found that it is an important oxide in chloride rich atmospheres. Haces et al.[2] propose that Cu is the main alloying element driving the formation of the protective layer in low-alloy steels.

Substituted magnetites have been studied by Mössbauer spectroscopy to see the changes in the hyperfine parameters[8-10] and particularly Cu-substituted magnetites have been studied by Bhaduri[11] for concentrations larger than 18%Cu. The role of Cu in the rust formation process, using the hydrolysis method, has been studied by Jaen et al.[7]. In the present study we have used the hydrolysis method to obtained Cu-substituted magnetites, with 1% and 5%Cu, to follow changes in the hyperfine parameters and also to see the role of Cu in its formation. The interesting point about using the hydrolysis method is the magnetite formation under the presence of sulfate ions, chloride ions and alloying elements simulating the corrosion process under atmospheric conditions. This study may be of interest when characterizing rust from weathering steels under continuous wet conditions and from chloride rich atmospheres.

2. Experimental

We have produced magnetite samples by the hydrothermal method of Schwertmann and Cornell[12], which basically consist in mixing an iron salt solution, in our case we used ferrous sulfate (FS) and ferrous chloride(FC), with a NaOH solution keeping the temperature constant and pumping N_2 during the experiment. 1% or 5% Fe, by weight, was substituted by Cu using a copper salt, cupric sulfate (CS) or cupric chloride (CC). A drawback of the method is the formation of magnetite with as a distribution of small particles with the possible oxidation to a nonstoichiometric product or to maghemite formation. Four samples were obtained starting with 100% FC, FC100(0%Cu), FCCC1(FC + CC,1% Cu), FCCC5(FC + CC,5%Cu) and FCCS5(FC + CS,5%Cu) another four samples were prepared based on FS, FS100, FSCC1, FSCC5, FSCS5. The samples were analyzed by transmission Mössbauer spectroscopy (MS) at room temperature and X-ray diffraction (XRD).

3. Results and discussion

Table 1 shows our results for the Mössbauer spectra(MS) of the magnetite samples, where HF, IS, QS, AREA, W stand for the hyperfine magnetic field, isomer shift (with respect to α-Fe), quadrupole splitting, relative areas and linewidths (outer lines) respectively. The fits to the experimental spectra were performed keeping the goethite IS and QS fixed to its standard values while leaving the other parameters free to change. For FS-magnetite four sextets were used to fit this MS, two for magnetite, one for crystalline goethite and the fourth one for poorly crystalline goethite which comes out very broad resembling a doublet. For FC-magnetite three sextets were used, the sextet for poor crystalline goethite is very broad looking as a doublet.
The hyperfine FS-magnetite parameters HF, QS, IS and W are insensitive to the 1%Cu presence. For 5%Cu substitution IS decreases and W increases for B-sites, the other parameters remaining as in pure magnetite. For the FC-magnetite W increases at B-sites with increasing Cu content more dramatically than for FS-magnetite and IS suffers a small increase for 5%Cu concentration. From the variation of W we can say that Cu substitute iron preferentially at B-sites.
The Mössbauer analysis shows the presence of goethite in all Cu-substituted samples in agreement with the results of the XRD measurements where only spinel and goethite reflections are seen and also a broadening of the peaks as the Cu concentration increases. From the MS the goethite has a component of good crystallinity together with a poor crystallinity component (α-FeOOH(p)) for FS-magnetite whereas it is of poor crystalline character for FC-magnetite. The amount of goethite increases with increasing Cu content and it is larger for FS-magnetite, this could be due to the presence of sulfate ions which according to studies of corrosion under atmospheres containing these ions the goethite production is favored[1,5]. It is

worth mentioning that goethite also appears during the formation of Al substituted magnetites by the hydrothermal method[13].

Fig. 1 shows the MS for Cu-substituted FS-magnetite including all subspectra. For FSCC1 we can clearly see the crystalline goethite peak. The peak at B-sites gets wider and its intensity decreases as the Cu concentration increases. For FC-magnetite spectra, not shown, the linewidth at B-sites increases but its relative intensity, as compared to the A-sites, remains almost the same as the Cu doping increases. We do not find a decrease of HF for FC-magnetite at both sites, similar to findings for low substituted Co magnetites where only the B-site linewidth changes[10]. For FS-magnetite the HF decreases slightly with increasing Cu content. These conflicting results may be due to difficulties with the small particle distribution[14] in our samples leading to spectra with highly overlapping sextets which may produce large uncertainties in HF values and other hyperfine parameters. To overcome these problems applied field Mössbauer spectroscopy will be used.

Table 1

Hyperfine parameters for magnetite samples (HF(KOe), IS(mm/s, α-Fe), QS(mm/s), AREA(%), W(mm/s))

Magnetite from FS						Magnetite from FC					
OXIDE	HF	IS	QS	AREA	W	OXIDE	HF	IS	QS	AREA	W
FS100						**FC100**					
Fe_3O_4	492	0.28	0.02	37.0	0.43	Fe_3O_4	489	0.29	0.01	44.6	0.42
	457	0.64	-0.03	63.0	0.64		455	0.64	-0.02	55.4	0.55
FSCC1						**FCCC1**					
Fe_3O_4	487	0.32	-0.01	30.3	0.39	Fe_3O_4	489	0.31	0.00	45.4	0.53
	456	0.66	0.01	40.7	0.61		454	0.60	-0.03	48.0	0.76
α-FeOOH	369	0.34	-0.27	9.2		α-FeOOH					
α-FeOOH(p)	325	0.34	-0.27	19.8		α-FeOOH(p)	43	0.34	.0.27	6.6	
FSCC5						**FCCC5**					
Fe_3O_4	487	0.31	-0.01	31.2	0.59	Fe_3O_4	488	0.44	0.02	27.6	0.49
	451	0.41	-0.03	24.9	0.80		454	0.72	-0.03	55.0	0.99
α-FeOOH	366	0.34	-0.27	27.4		α-FeOOH					
α-FeOOH(p)	126	0.34	-0.27	16.5		α-FeOOH(p)	92	0.34	-0.27	17.4	
FSCS5						**FCCS5**					
Fe_3O_4	488	0.35	-0.03	44.2	0.58	Fe_3O_4	489	0.45	-0.01	27.8	0.49
	450	0.51	-0.02	30.0	0.98		454	0.69	-0.02	50.5	1.06
α-FeOOH	380	0.34	-0.27	10.3		α-FeOOH					
α-FeOOH(p)	142	0.34	-0.27	15.5		α-FeOOH(p)	279	0.34	-0.27	21.7	

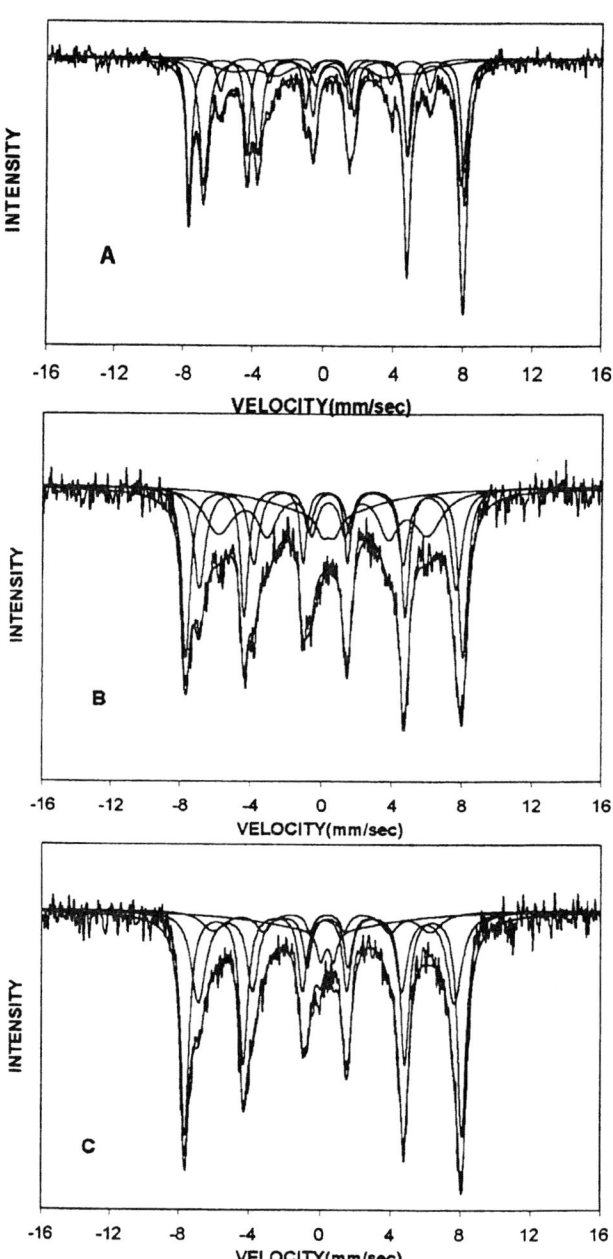

Fig. 1. Magnetite obtained from ferrous sulfate. A)
FSCC1, B) FSCC5, C) FSCS5

Conclusions

Our results may be of interest to analyze magnetite present in weathering steel rust produce under continuous wet conditions and under rich chloride atmospheres. Due to the large linewidth increase at B-sites we can conclude that Cu substitutes Fe preferentially there. The presence of copper also drives the formation of goethite, which could be an important mechanism for the formation of a protective rust layer in weathering steels. The sulfate ions involved in the process of magnetite formation seem also to favor goethite production as compared to chloride ions.

Acknowledgments

This work was partially supported by the Instituto Colombiano para el Desarrollo de la Ciencia y la Tecnología, Francisco José de Caldas, Colciencias. We thank Dr. Eddy De Grave for assistance in taking the X-ray spectra at the Department of Atomic and Radiation Physics, NUMAT Group, University of Gent, Belgium.

References

[1] J. Davalos, M. Gracia, J.F. Marco y J.R. Gancedo, Hyp. Int. 69 (1991) 871
[2] C. Haces, N.R. Furet and L. Muleshkova, Hyp. Int. 67 (1991) 587
[3] S. Nasrazadani and A. Raman, Corrosion 49 (1993) 294
[4] M. Yamashita, H. Nagano, T. Misawa and H.E. Townsend, Proc. of the 13th International Corrosion Conference, Melbourne, Australia, 1996.
[5] C. Arroyave and M. Morcillo, Trends in Corros. Sci. 2 (1997) 1.
[6] S.J. Oh, D.C. Cook and H.E. Townsend, Proc. of the International Conference on Applications of the Mössbauer Effect, Rio de Janeiro, Brazil, 1997.
[7] J.A. Jaen, J. Davalos and J.R. Gancedo, Proc. of the International Conference on Applications of the Mössbauer Effect, Rio de Janeiro, Brazil, 1997.
[8] E. De Grave, R.M. Persoons and R.E. Vandenbergue, Proc. of the XXXII Zakopane School of Physics, Zakopane, Poland, 1997.
[9] E. De Grave, R.M. Persoons, R.E. Vandenbergue and P.M.A. De Bakker, Phys. Rev. B 47 (1993) 5881
[10] G.M. Da Costa, E. De Grave, H. O'Neill and Ch. Laurent(submitted to J. Solid State Chem.)
[11] M. Bhaduri, J. Chem. Phys. 75 (1981) 3674.
[12] U. Schwertmann and R.M. Cornell, in Iron Oxides in the Laboratory(VCH, Weinheim, 1991) p. 111.
[13] U. Schwertmann and E. Murad, Clays and Clay Min. 38 (1990) 196
[14] G.M. da Costa, Ph.D. thesis, University of Gent, Belgium, 1996.

STUDYING THE ATMOSPHERIC CORROSION BEHAVIOR
OF WEATHERING STEELS AT A MILD MARINE ENVIRONMENT

Sei J. Oh[1], D.C. Cook[2], Soon-Ju Kwon[1] and H.E. Townsend[3]

[1] *Department of Materials Science and Engineering, Pohang University of Science and Technology, Pohang, Kyungbuk, 790-784, Korea.*

[2] *Department of Physics, Old Dominion University, Norfolk, VA 23529 U.S.A.*

[3] *Bethlehem Steel Corporation, Bethlehem, PA 18016 U.S.A.*

The atmospheric corrosion products on three types of weathering steels which were exposed at a marine environment for sixteen years, have been analyzed by Mössbauer spectroscopy and Electron Probe Micro-Analysis. In the region closer to the steel surface, silicon and chromium distributed with similar concentration to those in the steel substrate. Increasing the silicon content resulted in increasing the relative fraction of superparamagnetic goethite. Although the nickel content which gradually reduced toward the surface of the corrosion products was different, the composition of corrosion products was almost the same.

1. INTRODUCTION

Studying the atmospheric corrosion of steel is very complicated since there are many active parameters, such as composition of steel, environmental condition and exposure time[1-3]. Some studies have been performed to improve the understanding of the atmospheric corrosion behavior of steel as a function of the steel composition[4-5], in order to design steel with a better resistance against the atmospheric corrosion. The better corrosion resistance can be obtained from the formation of a protective layer which is densely packed with small-sized particles. It has been proposed in papers that the alloying elements would have important roles in forming the protective layer[6-7], although the mechanisms are not clearly resolved. This study aims to understand the atmospheric corrosion behavior of weathering steel exposed under a marine environment for the same exposure time. The better understanding can be obtained by correlating the Mössbauer and electron probe micro-analyzer data with the steel composition and the distribution of alloying elements in the corrosion products.

2. EXPERIMENTAL PROCEDURES

Table 1 lists the compositions of the three A588-type weathering steels, which were investigated in this study. Comparing to the contents of silicon and nickel of the coupon A52, the coupon A45 was lower in silicon, but the coupon A50 was lower in nickel. The coupons had been exposed for sixteen years at Kure Beach NC, USA, 250 m from the shoreline of the Atlantic Ocean. The atmospheric concentrations of chlorine and sulfur dioxides were 39 g/m^2 yr and 4.1 g/m^2 yr[4]. The Mössbauer spectroscopic samples were prepared from the corrosion products, which were scraped from the surface region of about 1x2 cm^2 on the steel substrate. The transmission spectra were recorded at 300K and 77K. After fitting the Mössbauer spectrum, the absorption area fraction of each iron oxide was converted to relative atomic fraction using the relative recoilless fractions[8]. The distribution of alloying elements across the intact corrosion layer was characterized by compositional mapping and line profile using an electron probe micro-analyzer(EPMA). The cross-sectional sample for the microanalysis was prepared by the conventional metallographic sample preparation method, i.e., mounting in epoxy, grinding and polishing. Although XRD data is not included in this paper, x-ray diffraction pattern was recorded, in order to identify whether or not akaganeite was present.

3.   RESULTS

The Mössbauer spectra of the corrosion products formed on the coupon A50 are shown in Fig. 1, where goethite (α-FeOOH), lepidocrocite (γ-FeOOH) and superparamagnetic maghemite (γ Fe_2O_3(s)) were identified using Mössbauer spectroscopy. No akaganeite was detected. The XRD analysis for the identification of akaganeite will be published soon. In the 300K Mössbauer spectrum, the superparamagnetic component of goethite (α-FeOOH(s)) which comprised the doublet together with lepidocrocite and superparamagnetic maghemite, were present. Goethite in Fig. 1(a) is classified as follows[9-10] : (1) α-FeOOH(m) : goethite particle of >15 nm, which is magnetic both at 300K and 77K, and (2) α-FeOOH(s) : goethite particle of <15 nm, which is superparamagnetic at 300K. The 300K subspectrum of magnetic goethite (α-FeOOH(m)) showed the distribution of hyperfine field with the most probable value of about 30 Tesla. Comparing to Cabral and Reyes' data[11], it was estimated that the mean particle size of goethite was approximately 20-30 nm. Also, it may be inferred that the contribution of the superparamagnetic goethite particles, <15 nm, were possibly present in the doublet.

Table 1. Composition of the three A588-type weathering steel coupons.

(Wt. %)

Coupon	C	Mn	P	S	Si	Ni	Cr	Cu	Al
A45	0 14	0 97	0.012	0 016	0 12	0 31	0 59	0 36	0.037
A50	0.12	1 06	0.006	0.013	0 77	0 01	0 59	0.33	0.038
A52	0.14	1 00	0.009	0 018	0 75	0.31	0 57	0.35	0.052

At 77K, α-FeOOH(s) is subclassified as follows[9-10] : (1) α-FeOOH(s1) : goethite particle of 8-15 nm, which is superparamagnetic at 300K, but magnetic at 77K, and (2) α-FeOOH(s2) : goethite particle of <8 nm, which is superparamagnetic at both temperatures. Using Kundig's data[12], the relative fraction of superparamagnetic goethite, α-FeOOH(s2), was estimated 8%. The Mössbauer spectra of the corrosion products on the other coupons, A45 and A52, were very similar to those in Fig. 1. Table 2 lists the iron oxide phases and their relative fractions identified in the corrosion products formed on the three weathering steel coupons. The relative fraction of each iron oxide was

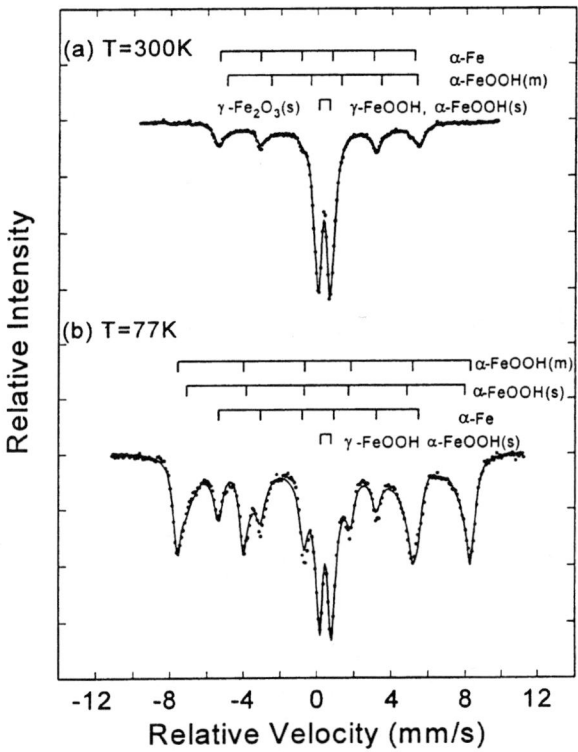

Fig. 1. Mössbauer spectra of corrosion products formed on coupon A50 exposed at the marine site of Kure Beach, NC for sixteen years.

Table 2. The relative fraction of iron oxides on the three coupons.

Coupon	Oxide Fraction (%) (error $\cdot$ ±1)						
	α-FeOOH (m)	α-FeOOH (s1)	α-FeOOH (s2)	α-FeOOH (s1+s2)	α-FeOOH (m+s1+s2)	γ-FeOOH	γ-Fe$_2$O$_3$(s)
A45	25	47	8	55	80	15	5
A50	21	51	8	59	80	15	5
A52	20	53	8	61	81	14	5

determined after subtracting the contribution of the steel powder, which was due to mechanically scraping from the steel substrate while the corrosion products were being removed.

The line scanning for studying the depth profiles of alloying elements was accomplished along the line in the middle of EPMA image, as shown in Fig. 2(a). The line scanning for probing the presence of chromium, nickel and silicon in the corrosion products is shown in Fig. 2(b). In the region of corrosion products closer to the steel surface, chromium and silicon concentrated with roughly similar density to those in the steel substrate, as shown in Fig. 3(b), where the larger dot density in Fig. 3(b) corresponds to the higher relative intensity in Fig. 2(b). The concentration did not appear in the region near the surface of the corrosion products. It is shown in Fig. 3(b) that silicon randomly aggregated as the form of small islands in the outer region of the corrosion coating contacting with

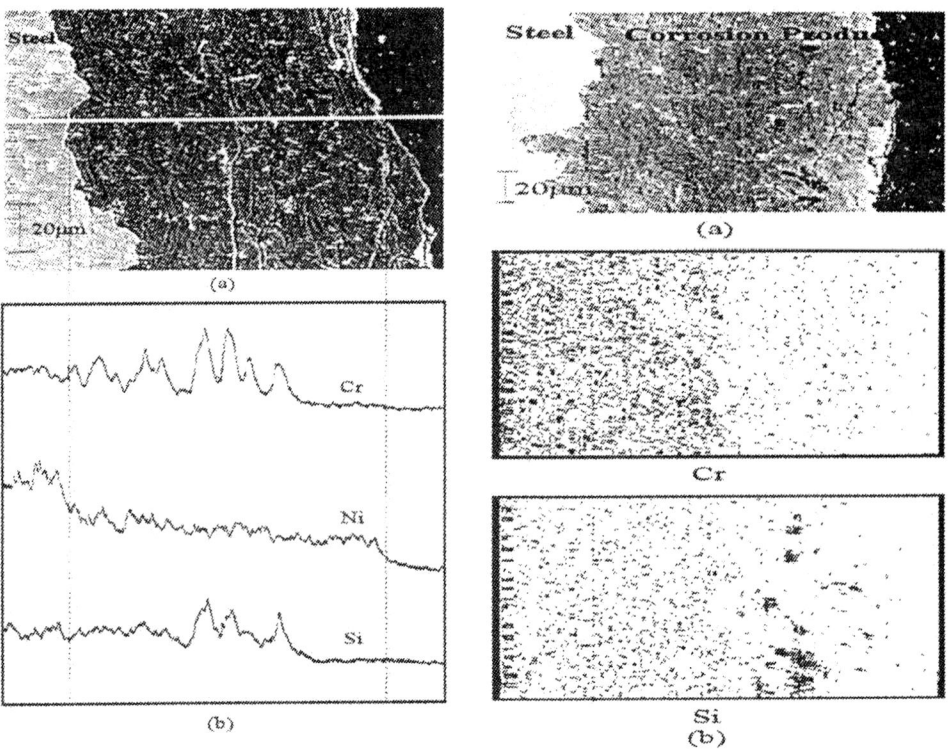

Fig. 2. EPMA image (a) and line profiles (b) of chromium, nickel and silicon in the corrosion products on the coupon A52 exposed at the marine site for sixteen years.

Fig. 3. EPMA image (a) and depth profiles (b) of chromium and silicon in the corrosion products on the coupon A50 exposed at the marine site for sixteen years

air, but not chromium. The nickel content gradually decreased toward the surface of the corrosion products formed on the coupons A45 and A52, as shown in Fig. 2(b), except for the coupon A50 due to the insignificant content of nickel in the steel substrate.

4.   DISCUSSION

The coupons A45 and A52 with different silicon content, Table 1, had almost the same relative fractions of constituent phases, i.e., goethite(m+s1+s2), lepidocrocite and superparamagnetic maghemite, as listed in Table 2. However, the relative fraction, 61%, of the small-sized goethite(s1+s2) on A52, which was higher in silicon content, was larger than that, 55%, on A45. It was indicated that increasing the silicon content in the steel substrate resulted in the increase the relative fraction of small-sized goethite particles exhibiting superparamagnetism. Increasing the fraction of small-size goethite could be helpful in the formation of the complete protective layer to prevent water and oxygen penetration to the steel substrate by densely compacting the small particles. Fig. 3(b) shows that the concentrations of silicon and chromium appeared to drop at the similar position in the corrosion products. Also, although the mapping of silicon and chromium on the coupon A52 do not appear in Fig. 3, it is revealed that their concentrations dropped at the similar position, as appeared on the coupon A50. However, although the chromium distribution was similar to others, the behavior of silicon in the coupon A45 could not be confirmed due to low concentration. The behaviors of chromium and silicon are interesting, since it may be inferred that they could have a certain interaction with iron oxides in the corrosion products, together. It was reported by Yamashita et al.[13] that increasing the chromium content in the synthetic chromium-substituted goethite ((α-Fe$_{1-x}$Cr$_x$)OOH) up to x=22wt.% resulted in the decrease of the goethite particle size from 90 nm for no chromium to 10 nm for greater than 3 wt.% chromium. However, it is not clear whether or not chromium and/or silicon, which diffused from the steel substrate, refined the goethite particles in the atmospheric corrosion products investigated in this study. The refining mechanism of iron oxide particles in atmospheric corrosion products will be a subject of further study.

Fig 2(b) shows that nickel content gradually decreased toward the surface of the corrosion products. The composition of corrosion products on the coupons A50 and A52 with different nickel content were very similar to each other in every aspect, i.e., the constituent phases, their relative fractions and particle size distribution, as listed in Table 2. An extensive research[4] revealed that increasing the nickel content in steel substrate resulted in decreasing corrosion rate of steel, which did not necessarily mean the particle size refinement. Unfortunately, the decrease of the corrosion rate with increasing nickel content could not be explained in this study.

5.   CONCLUSIONS

Nickel had negative gradient toward the surface of the corrosion products and had no detectable influence on the composition of corrosion products. In the region closer to the steel surface, silicon and chromium distributed with similar concentration to those in the steel substrate. Increasing the silicon content resulted in increasing the relative fraction of superparamagnetic goethite(s1+s2) in the corrosion products. The refining mechanism of iron oxide particles which formed in atmosphere, will be a subject of further study.

6.   ACKNOWLEDGMENTS

Supported by Old Dominion University graduate program and Korea Ministry of Education through postdoctoral fellowship program.

Reference

[1]   J. Davalos, M. Gracia, J.F. Marco and J.R. Gancedo, Hyp. Int. **69**, (1991) 871.

[2]   J.F. Marco, J. Davalos, M. Gracia and J.R. Gancedo, Hyp. Int. **57**, (1994) 606.

[3]   H. Kihira, S. Ito and T. Murata, Corr. Sci. **31**, (1990) 383.

[4]   C.R. Shastry, J.J. Friel and H.E. Townsend, ASTM STP 965 p. 5.

[5]   H.E. Townsend and J.C. Zoccola, ASTM STP 767 p. 45.

[6]   M. Yamashita, H. Miyuki, Y. Matsuda, H. Nahano and T. Misawa, Corr. Sci. **36**, (1993) 283.

[7]   M. Yamashita, H. Nagano, T. Misawa and H.E. Townsend, Proc. 13th Int. Congress on Corrosion, Clayton, Australia, 1996.

[8]   S.J. Oh. D.C. Cook and H.E. Townsend, Hyp. Int. **112**, (1997) 59.

[9]   L.H. Bowen, E.De Grave and R.E. Vandenberghe, in: Mössbauer Spectroscopy Applied to Magnetism and Materials Science Vol. 1, ed. G.J. Long (Plenum Press, New York, 1993).

[10]   E. Murad and J. H. Johnston in: Mössbauer Spectroscopy Applied to Inorganic Chemistry Vol. 2, ed. G.J. Long (Plenum Press, N. Y., 1987).

[11]   A. Cabral-Prieto and A. Reyes-Felipe, Proc. Int. Materials Research Congress, Cancun, Mexico, 1997.

[12]   W. Kundig, H. Bommel, G. Constabaris and R.H. Lindquist, Phys. Rev. **142**, (1966) 327.

[13]   M. Yamashita, H. Miyuki, H. Nahano and T. Misawa, Corr. Eng. **43**, (1994) 43.

SYNTHESIS OF GOETHITE FROM SULPHATE AND CHLORIDE IRON *SOLUTIONS* IN THE PRESENCE OF Cu^{2+} AND Mn^{2+} IONS: CHARACTERISATION BY MÖSSBAUER SPECTROSCOPY

C. Arroyave[1], G. Perez[2] and A. L. Morales[1,2]

[1] Grupo de Corrosión y Protección. Departamento de Ingeniería Metalúrgica y de Materiales, Universidad de Antioquia. A.A 1226, Medellín, Colombia.
[2]Departamento de Física, Universidad de Antioquia, A.A. 1226, Medellín, Colombia.

Abstract

Looking for the assessment of the effect of exogen and endogen substances on the rust characteristics, several goethite samples were obtained by hydrothermal synthesis, with Cl$^-$ and SO$_4^{2-}$ as precursor ions, simulating the presence of the major atmospheric pollutants. Also, Cu and Mn were included as dopant elements, simulating the presence of two common alloying elements in weathering steel. As a main conclusion, the results indicate that the presence of Cu and Mn has a significant effect on the crystallity and/or particle size of the synthetised goethites what is reflected in a drastic decrease of the relative intensity of the magnetic component in the Mössbauer spectra.

1. Introduction

It has been stated that the presence of SO$_2$ as atmospheric pollutant plays a decisive role in the formation of protective corrosion films on weathering steel[1]. In atmospheric corrosion the action combined of chloride and sulfate ions may decrease the corrosion rate [2].

The composition of rust layers formed on the surface of weathering steel in natural conditions depends upon the steel copper content and the exposure time, moreover, the lower corrosion rate of weathering steel is a consequence of the effect of copper in the formation of a higher concentration of superparamagnetic goethite[3].

Natural goethite contains substitutional elements such as Al [4], Cu and Mn, and presents a Mössbauer spectra with the characteristic of small particles. The observations about synthetic and substituted goethites, with Cu and Mn, could be used as a model in understanding the formation of the protective rust layer on the surface of weathering steel. With this goal in mind, it has been started a Mössbauer study of goethites synthetised in the presence of Cu^{2+} and Mn^{2+} ions..

2. Experimental

Sample preparation. The samples GsF50cF50, GsF75cF25, GsF100 , GcF100, have been prepared according to the method of U. Schwertmann and R. M. Cornell [5], where 9.9 g of unoxidized cristal FeCl$_2$.4H$_2$O or 13.9 g FeSO$_4$.7H$_2$O are dissolved in 1 l distilled water, then 110 mL solution 1M NaHCO$_3$ is added .Oxidation is complete within 72 hours. The pH during oxidation is self-controlled to about 7.

The synthetised samples in the presence of Cu^{2+} ions (GcFcC5, GcFcC1 y GsFcC1), and Mn^{2+} ions(GsFcM5,GcFcM5 y GcFsM5), in principle followed the same method of preparation, except that the iron solution is replaced by a mixture of Fe-Cu or Fe-Mn solution. By varying the quotient Mn/[Mn + Fe] or Cu/[Cu + Fe], different grades of cations can be obtained. Table 1 describes the synthesis for each sample.

Table 1. Solutions Used for the Goethite Synthesis

Sample	synthesis
GcF100	$FeCL_2$ + $NaHCO_3$
GcFcM5	$FeCL_2$ + 5%$MnCL_2$ + $NaHCO_3$
GcFsM5	$FeCL_2$ + 5%$MnSO_4$ + $NaHCO_3$
GcFcC1	$FeCL_2$ + 1%$CuCL_2$ + $NaHCO_3$
GcFcC5	$FeCL_2$ + 5%$CuCL_2$ + $NaHCO_3$
GsF100	$FeSO_4$ + $NaHCO_3$
GsF75cF25	75%$FeSO_4$ + 25% $FeCL_2$ + $NaHCO_3$
GsF50cF50	50%$FeSO_4$ + 50% $FeCL_2$ + $NaHCO_3$
GsFcM5	$FeSO_4$ + 5% $MnCL_2$ + $NaHCO_3$
GsFcC1	$FeSO_4$ + 1% $CuCL_2$ + $NaHCO_3$

X-Ray Diffraction (XRD). XRD measurements were perfomed on our samples with a RIGAKU Miniflex 2005 diffractometer equipped with a Cu(Kα) cathode and Ni filter. The scans were done in the range of 5° to 60° (2θ) at a speed of 2° per minute.

Mössbauer Spectroscopy. Mössbauer spectra were collected at room temperature, with a time–mode spectrometer using a constant acceleration drive and triangular reference signal. Velocity calibration was regularly achieved by taking spectra of standard hematite. All absorbers were prepared with a thickness of about 12,5 mgcm^{-2}a α-Fe.

For Mössbauer spectral analysis two computer programs were used: **MOSF** and **DIST3E** [6]. The **MOSF** program is based on a nonlinear least-squares fitting procedure assuming Lorentzian line shapes, and the **DIST3E** program which is based on a distribution of hyperfine fields and/or quadrupole splittings.

3. Results and discussion

X Ray Diffraction. For the samples that were made from sulfates, the XRD shows both goethite and lepidocrocite peaks. The samples that were obtained from chlorides, and also those doped with copper or manganese , show only goethite peaks.

Mössbauer Spectra. All room temperature Mössbauer spectra, Figure 1, show a central doublet and a hyperfine magnetic component (sextet) with broad lines. In this figure spectra showing subspectra components were fitted with program **MOSF** including GcFcC5. For samples obtained from sulfates, the central doublet corresponds to a mixture of lepidocrocite and superparamagnetic goethite. The spectrum corresponding to the sample synthesised from a solution containing 5% $CuCl_2$ shows only one doublet. This doublet may be ascribed either to superparamagnetic goethite, particles of very poor cristallinity or to copper presence in the goethite structure. The respective hyperfine field distributions for samples are show in figure 2. Those samples formed from sulfates, synthesised in presence of Cu^{2+} and Mn^{2+} ions, are more crystalline than those obtained from chlorides in the presence of Cu^{2+} and Mn^{2+} ions, due to the presence of a larger sextet component.

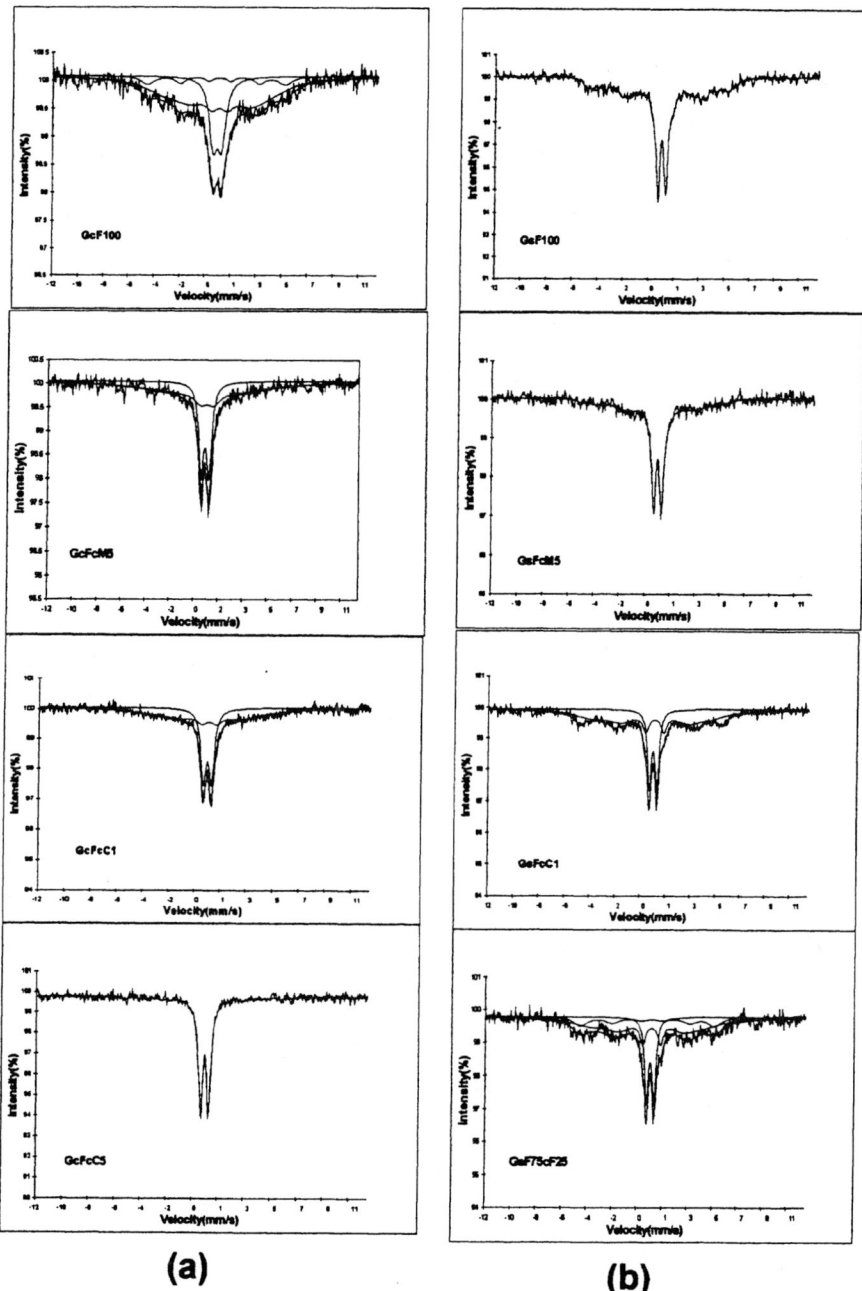

Figure 1: Mossbauer Spectra of synthetic samples: (a) From
chloride iron salts and (b) From sulfate iron salts.

C. Arroyave et al. / Synthesis of goethite from sulphate and chloride iron solutions in the presence of Cu^{2+} and Mn^{2+} ions: characterisation by Mössbauer spectroscopy

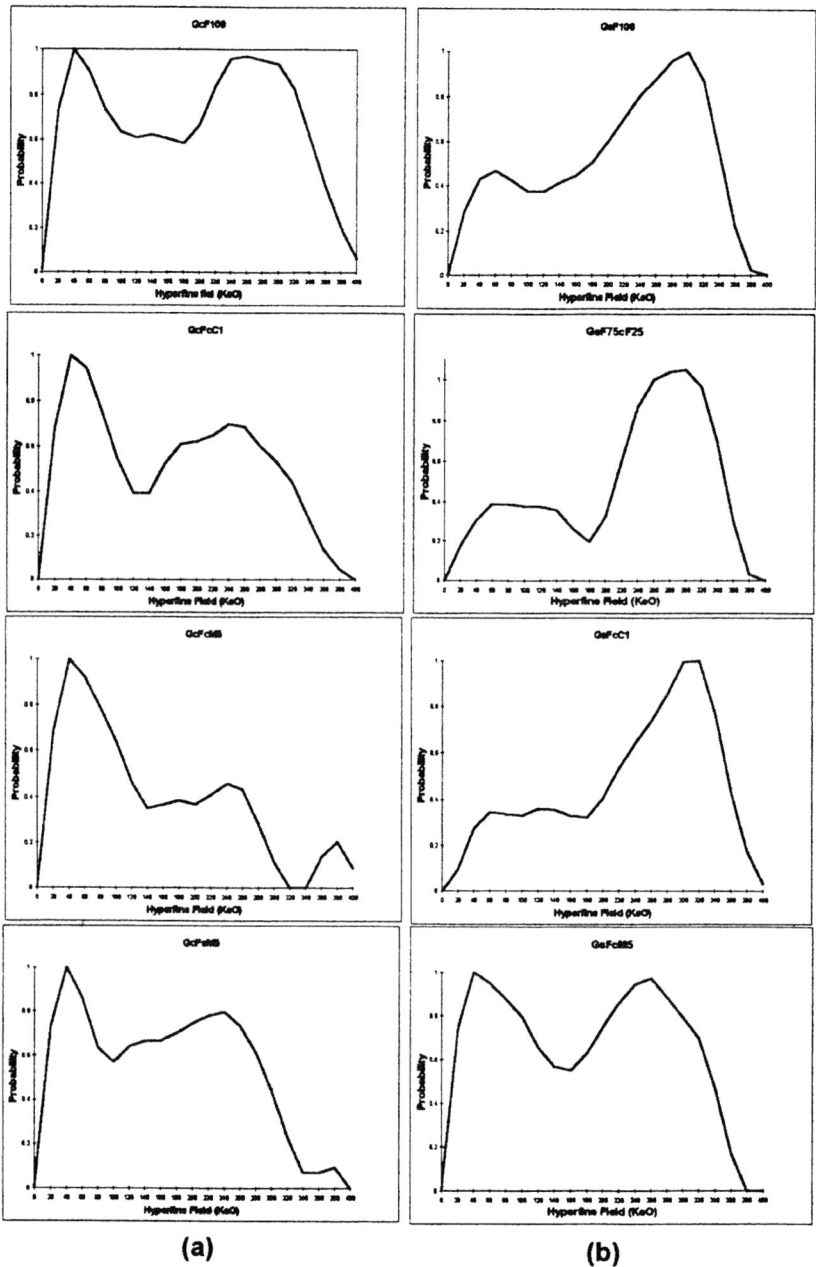

Figure 2: Hyperfine Field Distribution of Synthetic samples: (a) From chloride iron salts and (b) From sulfate iron salts

Table 2. Mössbauer Parameters of the Samples with the DIST3E Program

Sample	Comp.	H_{Max}	$\overline{H}$	Q_{Max}	$\overline{Q}$	IS	Area(%)
GcF100	Sextet	45	191			0.34	86
	Doublet			0.53	0.50	0.34	14
GcFcM5	Sextet	46	139	.		0.34	57
	Doublet			0.50	0.58	0.34	43
GcFsM5	Sextet	43	161			0.34	74
	Doublet			0.53	0.58	0.34	26
GcFcC1	Sextet	47	167			0.34	60
	Doublet			0.50	0.58	0.34	40
GsF100	Sextet	294	213			0.34	69
	Doublet		.	0.54	0.55	0.37	31
GsF75cF25	Sextet	292	228	0.51	0.55	0.34	72
	Doublet					0.38	28
GsF50cF50	Sextet	288	231			0.34	53
	Doublet			0.53	0.56	0.38	47
GsFcM5	Sextet	46	178			0.34	58
	Doublet			0.52	0.55	0.37	42
GsFcC1	Sextet	310	235			0.34	69
	Doublet			0.55	0.54	0.43	31

H_{max}: Hyperfine field of maximum probability in KOe, $\overline{H}$: Average hyperfine field in KOe, Q_{max}: Quadrupole splitting of maximum probability in mms⁻¹, $\overline{Q}$:Average quadrupole splitting in mms⁻¹ and IS:Isomer shift in mms⁻¹, referred to α-Fe.

Table 3. Mössbauer parameters of the samples with the MOSF Program.

Sample	Comp.	H(KOe)	Q(mms⁻¹)	IS(mms⁻¹)	Area(%)
GcF100	Sextet	308	-0.27	0.34	11
	Sextet	218	-0.27	0.34	64
	Doublet		0.60	0.34	25
GcFcM5	Sextet	125	-0.27	0.34	60
	Doublet		0.59	0.34	40
GcFsM5	Sextet	193	-0.27	0.34	68
	Doublet		0.61	0.34	32
GcFcC1	Sextet	200	-0.27	0.34	65
	Doublet		0.55	0.34	35
GcFcC5	Doublet		0.58	0.34.	100
GsF100	Sextet	305	-0.27	0.34	13
	Sextet	222	-0.27	0.34	59
	Doublet		0.54	0.37	28
GsF75cF25	Sextet	301	-0.27	0.34	19
	Sextet	250	-0.27	0.34	56
	Doublet		0.53	0.38	25
GsF50cF50	Sextet	305	-0.27	0.34	26
	Sextet	210	0.55	0.34	30
	Doublet			0.38	44
GsFcM5	Sextet	216	-0.27	0.34	60
	Doublet		0.55	0.37	40
GsFcC1	Sextet	270	-0.27	0.34	67
	Doublet		0.53	0.38	33

H: Hyperfine field, Q: Quadrupole shift, IS:Isomer shift in mms⁻¹, referred to α-Fe

The hyperfine parameters obtained using **DIST3E** are listed in table 2. For the hyperfine magnetic field we give the most probable value H_{max} and the average value $\bar{H}$. The last one is larger for samples produced from sulfate salts due to the presence of more crystalline particles.

In table 3 the hyperfine parameters obtained from **MOSF** are reported for comparison, although **DIST3E** gives better results[4] in this case. The hyperfine magnetic field parameters from **MOSF** are very similar to the average values obtained with **DIST3E**.

4. Conclusions

The procedure for obtaining pure goethite from chloride iron salts, with a magnetic component in the range 72-80%, is good and reproducible. For samples obtained from sulfate iron salts a mixture of goethite and lepidocrocite is found. The influence of sulfate ions on the goethite formation is to produce a better cristallinity but with the drawback of forming lepidocrocite.

The presence of Mn in goethite results in a reduction of the magnetic hyperfine field which is smaller than in the case of Cu. The Cu and Mn presence favor the formation of either superparamagnetic goethite, poor crystalline gothite particles or to copper presence in the goethite structure.

According to the literature[1,3,7] the presence of superparamagnetic goethite helps to form a protective layer in weathering steels. The present study shows that a similar process may occur in weathering steels. The results can be used to understand the process of atmospheric corrosion and the effect of Cu and Mn on the rusting of weathering steel.

Acknowledgements
We thank the NUMAT group at the Department of Atomic and Radiation Physics, University of Gent, Belgium; for providing the programs MOSF and DIST3E.

References
[1] J. F. Marco, J. Dávalos, M. Gracia and J. R.Gancedo, Hyp. Int., 57 (1990) 1991 – 1996.

[2] A.L. Morales et al., Caracterización de la corrosión del acero en presencia de cloruro y sulfatos, Revista colombiana de física, 29, (1997).

[3] C. Haces, N.R. Furet and Muleshkova, Hyp. Int., 67 (1991) 587-594.

[4] C.Barrero et al., Un Estudio Mössbauer Sobre Goethitas Sinteticas: Primero Resultados, Revista colombiana de física, 27, (1995).

[5] U.Schwertmann and R. M. Cornell, in Iron Oxides in the Laboratory(VCH, WEINHEIM, 1991) p. 28.

[6] R. Vandenberghe, E. De grave, and P.M.A. de Bakker, Hyp. Int., 83 (1994)

[7] S.J. Oh. D.C. Cook and H.E. Townsend, Proc. Of the ICAME, Rio de Janeiro, Brazil,1997.

STABILITY OF OXIDE MICROCRYSTALS IN Fe-ZEOLITE L

S. G. Marchetti, A. M. Alvarez, J. F. Bengoa, M. V. Cagnoli, N. G. Gallegos, A. A. Yeramián
CINDECA, CONICET, CIC, Facultad de Ciencias Exactas, Facultad de Ingeniería Universidad
Nacional de La Plata, 47 N° 257, 1900 La Plata, Argentina
R. C. Mercader
Departamento de Física, Universidad Nacional de La Plata,
C. C. 67, 1900 La Plata, Argentina.

Abstract

We have found by Mössbauer spectroscopy that the same iron species are present in systems of iron oxide supported on both potassic, ZLK and acidic, ZLH, zeolite L, but with different particle size distributions. The α-Fe_2O_3 microcrystals are located preferentially inside the zeolite channels in Fe/ZLK but not in Fe/ZLH. Water removal by outgassing in Fe/ZLH produces a breaking of the external α-Fe_2O_3 crystallites and a decrease of their average diameter. Instead, in Fe/ZLK it leads to the migration of the smaller crystals out of the pore structure and to a sintering of the particles.

Introduction

Molecular sieves are being used currently with growing interest as support of oxide and metallic catalysts because they can display shape selectivity in some reactions. Inside the zeolite channels the metallic particles have sizes that do not exceed *ca.* 1 to 2 nm and this allows to modify the catalyst selectivity and other properties. In the case of Fe-exchanged ZSM-5 and Y-zeolites it is known [1] that iron goes into the zeolite pore structure after impregnation. However, when these systems are subjected to reduction treatments in order to obtain the desired uniform-dispersed metal particles inside the zeolite structure, Fe migrates out of the pores [1] by mechanisms which are not yet fully understood [2]. When this happens, the beneficial effect of the zeolite channel structure no longer influences the outcome of the reaction. When considering the design of new Fe-zeolite catalysts it is of great importance to investigate the reason why this migration occurs and how it can be avoided. In the present work, Mössbauer spectroscopy is used to investigate if the above described behaviour is also observed in the Fe-zeolite L system.

Experimental

Potassic (ZLK) and acidic (ZLH) zeolite-L were used as supports for iron oxide phases. The former is the commercial form (Tosoh Corp.) of the zeolite-L. ZLH was obtained from ZLK after exchanging with a NH_4NO_3 (0.3 M) solution during 5h at 368 K for three times and calcination in air at 873 K for 22 h. The solids were prepared by dry impregnation of the zeolites with a solution of $Fe(NO_3)_3.9H_2O$ at a concentration high enough to yield solids with *ca.* 5% w/w of iron. The samples, in the following denoted as Fe/ZLK and Fe/ZLH, were dried in air and subsequently calcinated in dry N_2 stream from room temperature up to 698 K at a heating rate of 0.33 °/min, and kept at 698 K for 8 h. The solids were characterised by atomic absorption spectroscopy to determine the total iron content, by N_2 adsorption to measure the specific area, and by Mössbauer spectroscopy in air at 298 and 15 K.

After calcination, Fe/ZLK and Fe/ZLH were evacuated to 10^{-6} torr and kept for 9 h at 423 K and subsequently for 1 h at 673 K in a cell that allows to take "in situ" Mössbauer spectra [3]. After outgassing, the cell was filled with ultra-high-pure He and Mössbauer spectra of these samples, in the following Fe/ZLK(vac) and Fe/ZLH(vac), were registered at 298 and 15 K.

The spectra were evaluated by using a non-linear least-squares fitting programme with constraints. All components of each hyperfine split signal have the same Lorentzian linewidths. The spectra with very broad lines were simulated with a programme [4] capable to analyse hyperfine parameters distributions. All isomer shifts are referred to α-Fe at room temperature.

Thermo-gravimetric analysis (TGA) assays on ZLK and ZLH were carried out to study the elimination and uptake of water. Three cycles of heating up to 923 K and cooling in He stream (*ca.* 10 ppm of H_2O) without previous treatment were realised. Afterwards the supports remained for 24 h in air at room temperature and the weight increase was registered.

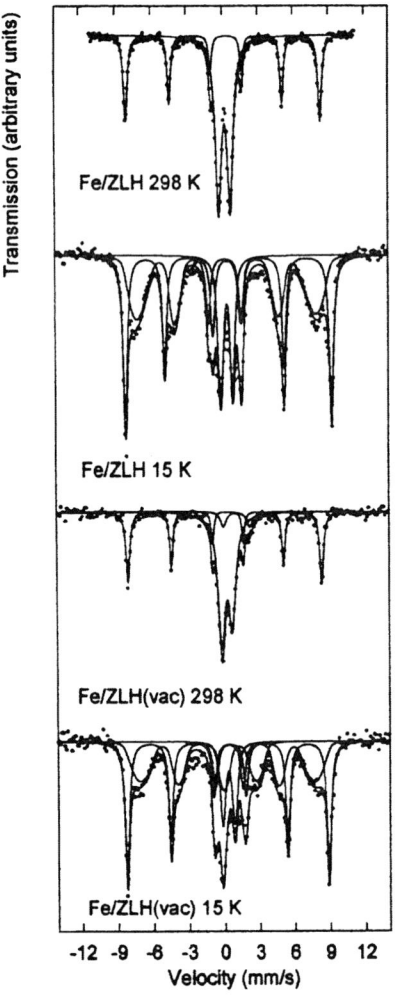

Figure 1: Mössbauer spectra of Fe/ZLH and Fe/ZLH (vac) at indicated temperatures.

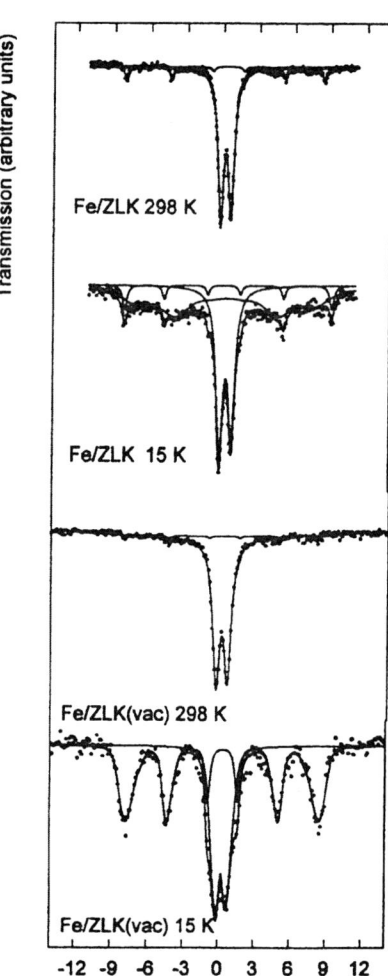

Figure 2: Mössbauer spectra of Fe/ZLK and Fe/ZLK(vac) at indicated temperatures.

Results and Discussion

The Mössbauer spectra at 298 and 15 K of the samples are shown in Figures 1 and 2. Fe/ZLH and Fe/ZLK display two signals at RT: a magnetic sextet and a paramagnetic doublet. The sextet is more intense for Fe/ZLH. When the temperature is lowered to 15 K, a second magnetic signal is observed in Fe/ZLH. Instead, in Fe/ZLK only two resolved signals are still noticed, but the background is significantly curved.

TABLE I: Mössbauer hyperfine parameters.

Temp.	Species	Parameters	Fe/ZLH	Fe/ZLH(vac)	Fe/ZLK	Fe/ZLK(vac)
298 K	$\alpha\text{-}Fe_2O_3$	H(T)	50.9 ± 0.1	50.9 ± 0.1	51.2 ± 0.1	49.9 ± 0.3
		δ(mm/s)	0.36 ± 0.01	0.35 ± 0.01	0.38 ± 0.02	0.36 ± 0.03
		2ε(mm/s)	-0.23 ± 0.01	-0.23 ± 0.01	-0.24 ± 0.03	-0.25 ± 0.07
		Γ(mm/s)	0.29 ± 0.01	0.26 ± 0.01	0.34 ± 0.01	0.26 ± 0.10
	Fe^{3+}	δ(mm/s)	0.34 ± 0.01	0.37 ± 0.01	0.32 ± 0.01	0.31 ± 0.01
		Δ(mm/s)	0.97 ± 0.01	0.89 ± 0.02	0.87 ± 0.01	0.96 ± 0.01
		Γ(mm/s)	0.58 ± 0.01	0.72 ± 0.03	0.57 ± 0.01	0.73 ± 0.01
	Fe^{2+}	δ(mm/s)	—	1.20 ± 0.06	—	—
		Δ(mm/s)	—	2.23 ± 0.14	—	—
		Γ(mm/s)	—	0.58 ± 0.15	—	—
15 K	$\alpha\text{-}Fe_2O_3$	H(T)	54.0 ± 0.1	53.0 ± 0.1	53.7 ± 0.1	—
		δ(mm/s)	0.48 ± 0.01	0.48 ± 0.01	0.45 ± 0.02	—
		2ε(mm/s)	0.34 ± 0.01	-0.12 ± 0.01	0.36 ± 0.03	—
		Γ(mm/s)	0.44 ± 0.01	0.44 ± 0.02	$0.34 \pm 0.00^{\bullet}$	—
	$\alpha\text{-}Fe_2O_3$	H(T)	46.3 ± 0.01	45.5 ± 0.2	$46.3 \pm 0.0^{\bullet}$	49.5 ± 0.2
		δ(mm/s)	0.47 ± 0.01	0.43 ± 0.02	$0.47 \pm 0.00^{\bullet}$	0.47 ± 0.02
		2ε (mm/s)	-0.01 ± 0.03	-0.11 ± 0.04	$-0.01 \pm 0.00^{\bullet}$	$0.00 \pm 0.00^{\bullet}$
		Γ(mm/s)	1.50 ± 0.10	1.32 ± 0.07	2.99 ± 0.22	1.13 ± 0.09
	Fe^{3+}	δ(mm/s)	0.45 ± 0.02	0.44 ± 0.02	0.41 ± 0.01	0.38 ± 0.01
		Δ(mm/s)	1.00 ± 0.02	1.02 ± 0.05	1.01 ± 0.01	1.07 ± 0.02
		Γ(mm/s)	0.96 ± 0.03	0.96 ± 0.07	0.69 ± 0.01	1.01 ± 0.07
	Fe^{2+}	δ(mm/s)	—	1.40 ± 0.05	—	—
		Δ(mm/s)	—	2.75 ± 0.10	—	—
		Γ(mm/s)	—	0.66 ± 0.11	—	—

*Parameter held fixed while fitting.

The Mössbauer spectra of Fe/ZLH at 298 and 15 K have hyperfine parameters (see Table I) that can be assigned to two iron species, $\alpha\text{-}Fe_2O_3$, and Fe^{3+} ions (probably exchanged with the support). The second sextet, evident at 15 K, has parameters similar to those of small hematite particles that have not experienced the Morin transition. Probably, this fraction of crystallites has not completed its magnetic splitting at 15 K and is still in a relaxation regime that partially gives raise to the broad lines. The $2\varepsilon \approx 0.0$ mm/s, and the very broad lines are characteristic of very small particles with a large size distribution. To estimate roughly their average size we applied the Collective Magnetic Excitation Model (CMEM) [5] assuming that all the magnetic hyperfine field diminution is originated in the reduced crystallite size. According to Vandenberghe [6] we take $H(V\infty)_{15K} = 53.2T$ as the saturation value for the field of hematite particles that are still in a weak-ferromagnetic state at

80 K with $2\varepsilon \approx 0.0$ mm/s, and applying the CMEM we estimate an average size of 3.2 nm. Despite this very small size, these crystals cannot be located inside de 1.2 nm channels of the support.

The sharper magnetic sextet, that at 15 K shows a quadrupole shift of 0.34 mm/s, reveals the existence of a fraction of larger α-Fe_2O_3 particles that undergo the Morin transition and, therefore, have an average diameter ≥ 20 nm [6]. This estimation is almost coincident with the CMEM result, that yields an average diameter of 17 nm, and indicates that, for this sample, the α-Fe_2O_3 particles lie preferentially outside the channel structure of the support.

TABLE II: Iron loading and specific area of the samples

	ZLK	Fe/ZLK	ZLH	Fe/ZLH
S_g (m^2/g)	290	33	345	66
Fe % (w/w)	—	5.84	—	4.95

The central signal that can still be observed at 15 K, might arise from the contribution of Fe^{3+} ions exchanged with H^+ of the support during the impregnation step, and also from very small α-Fe_2O_3 particles with a total superparamagnetic relaxation at this temperature. Considering the sample thermal pre-treatment, according to Fitch and Rees [7], the exchanged ions must be located at the A sites of the hexagonal prisms of the zeolite-L with a highly asymmetrical enviroment as deduced from the large Δ value found.

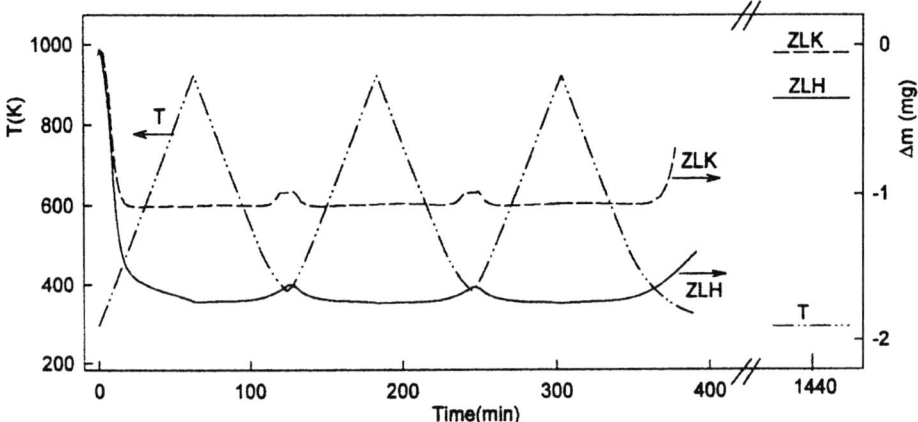

Figure 3: Three cycles of TGA of ZLH and ZLK in He flow (10 ppm of water). The weight loss of ZLH and ZLK after 24 h of air exposition at room temperature is shown on the right.

The Mössbauer spectrum of Fe/ZLK at RT is similar to that of Fe/ZLH but with a central doublet that displays (Figure 2) a larger relative area (89 ± 3 % for Fe/ZLK and 61 ± 2 % for Fe/ZLH). Its hyperfine parameters (Table I) are essentially the same as Fe/ZLH. This similarity no longer holds at 15 K. Instead of the second sextet of Fe/ZLH (Figure 1) the spectrum displays a curved background probably originated in a fraction of small particles undergoing an incomplete magnetic splitting. The fitting was simulated with one sextet, one doublet and a second sextet of very broad lines. The hyperfine parameters are shown in Table I. The relative area of the sharper sextet (11 ± 2%) is the same (within experimental errors) at RT and at 15 K. Its quadrupole shift is 0.34 mm/s at 15

K, denoting that this fraction of hematite particles undergoes the Morin transition and therefore its average diameter is ≥ 20 nm. Applying CMEM as above, a diameter of 20 nm is obtained. Although the fitting procedure is a rough approximation to the physical process actually taking place, the method yields an estimate of the fraction of the particles in the relaxing magnetic regime (55 ± 7 %). Because these particles at 15 K have not reached the degree of magnetic order of the Fe/ZLH particles, their size must be even smaller than 3.2 nm. In addition, the central doublet of Fe/ZLK, at 15 K, has a relative area larger than for Fe/ZLH (38 ± 2 % vs. 17 ± 1 %, respectively). We assign this increase to the presence of very small α-Fe_2O_3 superparamagnetic particles, since it is expected that the fraction of Fe^{3+} exchanged ions should be similar for both supports. The existence of this small particles fraction is coherent with our previous results [8], where we found that the K^+ exchanged by Fe^{3+} produces an increase of the pH of the solution that fills the channels of the zeolite so causing the precipitation and "anchoring" of ferrhydrite and/or goethite. After calcination, these species would generate the smallest α-Fe_2O_3 microcrystals.

The Mössbauer spectra of Fe/ZLH(vac) at 298 and 15 K are also shown in Figure 1. Both were fitted in the same way as Fe/ZLH but an additional doublet, assignable to Fe^{2+}, was needed. Relative to Fe/ZLH, the Fe/ZLH(vac) parameters (Table I) denote that both hematite sextets are originated in particles of smaller average diameters. The crystal size reduction of both fractions may be explained if some of the iron oxide crystals of this sample are envisaged obstructing the pore mouths as indeed is indicated by the strong decrease in the specific surface area measured after impregnation (Table II). Considering that the zeolite channels are straight and cylindrical, the presence of these crystallites would exclude a large part of the internal area from being monitored by BET. When the outgassing occurs, the water vapour would generate high pressures at microscopic levels inside the closed pores, breaking the crystals.

The hyperfine parameters of the Fe^{2+} signal are in agreement with those found by Fitch and Rees [7] for Fe^{2+} located at A sites. The Fe^{3+} reduction, produced by the outgassing, could be due to the dehydroxilation of the zeolite, as our TGA results show (see below). Similar results have been found by other authors in Fe-zeolite L [9] and Fe-zeolite X [10] systems.

Two signals can be seen in the Mössbauer spectrum of Fe/ZLK(vac) (Figure 2) at RT: a central doublet and a hardly noticeable sextet (4 ± 2 %). The parameters of this weak magnetic signal can be assigned to α-Fe_2O_3 (Table I) with its hyperfine field diminished. When the temperature decreases to 15 K, a broad and intense sextet is evident in addition to the still important central doublet. The weak and narrow magnetic signal cannot be resolved from the broad sextet. The Mössbauer parameters (Table I) of the broad sextet are characteristic of very small α-Fe_2O_3 crystallites. Using the CMEM we estimate an average crystal diameter of about 4 nm for this fraction. These findings are coherent with a picture where the larger crystals decrease their size for similar reasons than for Fe/ZLH(vac), while the smaller ones migrate out of the zeolite channels and sinter at the external surface promoted by water vapour.

Our TGA results show that the complete water elimination in ZLK occurs when the temperature of 494 K is surpassed (Figure 3) with the highest rate of water loss at 374 K (estimated from the slope of the loss of weight vs. time curve). However, if the solid is cooled back below 494 K the water is quickly reabsorbed. We infer that after calcination at 698 K during the sample preparation, water is completely eliminated from the solid but is regained as soon as it is in contact with air at room temperature. This is demonstrated by the sample returning to its original weight (see Figure 3). Differently, the acidic sites presence in ZLH makes that the water loss by dehydroxilation continues even at 923 K and the solid does not recover its original weight after it is exposed to air at room temperature. This irreversible loss of water would explain the Fe^{2+} ions presence in Fe/ZLH(vac) but not in Fe/ZLK(vac).

Conclusions

After calcination in air at 698 K for 8 hours, the α-Fe_2O_3 crystals get inside the zeolite channels in a higher proportion for Fe/ZLK than for Fe/ZLH. However, it has been demonstrated that the sudden removal of adsorbed water (heating in vacuum at a higher temperature than that of the highest rate of water loss) produces the crystal migration and subsequent sintering. We conclude that to prevent the exit of the crystals located inside the channels of the zeolite, it is very important to exercise care during the preparation steps, especially with the speed of water removal.

Acknowledgements

The authors acknowledge support of this work by Consejo Nacional de Investigaciones Científicas y Técnicas (PIP 4326-653/97, PEI 0507/97 and 0508/97), ANPCyT (PIP 1277-047/98), Comisión de Investigaciones Científicas Pcia. Bs. As. and Universidad Nacional de La Plata, Argentina. They wish also to thank Dr. E. Ponzi for recording the TGA spectra.

S.G.M., M.V.C., N.G.G., R.C.M. and A.A.Y. are members of Carrera del Investigador Científico y Tecnológico, CONICET. A.M.A. and J.F.B. are members of Carrera del Personal de Apoyo, CONICET and CIC, respectively.

References

[1].-T. Lin, Northwestern University, Ph. D., University Microfilms International, 300 N. Zeeb Road, Ann Arbor, MI 48106 (1984).

[2].-P. A. Jacobs in: Metal Microstructures in Zeolites. Preparation-Properties-Applications. Studies in Surface Science and Catalysis, Vol.12, eds. P. A. Jacobs, N. Y. Jaeger, P. Jíru and G. Schulz-Ekloff (Elsevier, Amsterdam, 1982) p.71.

[3].-S. G. Marchetti, J. F. Bengoa, M. V. Cagnoli, A. M. Alvarez, N. G. Gallegos, A. A. Yeramián and R. C. Mercader, Meas. Sci. Tech. 7 (1996) 758.

[4].-R. E. Vandenberghe, P.M.A. de Bakker and E. De Grave, Hyperfine Interact. 83, 29 (1994).

[5].-S. Mørup and H. Topsøe, Appl. Phys. 11 (1976) 63.

[6].-R. E. Vandenberghe in: Mössbauer Spectroscopy and Applications in Geology, International Training Center for Post-Graduate Soil Scientists (Belgium, 1991).p.4-6.

[7].-F. R. Fitch and L. V. C. Rees, Zeolites 2 (1982) 33.

[8].-S. G. Marchetti, A. M. Alvarez, J. F. Bengoa, M. V. Cagnoli, N. G. Gallegos, R. C. Mercader and A. A. Yeramián, Hyperfine Interact. C. 3 (1998) 77.

[9].-R. W. J. Wedd, B. V. Liengme, J. C. Scott and J. R. Sams, Solid State Comm. 7 (1969) 1091.

[10].-J. A. Morice and L. V. C. Rees, Trans. Faraday Soc. 64 (1968) 1388.

VIBRATIONAL PROPERTIES FROM MÖSSBAUER SPECTROSCOPY

A. M. Khasanov and J. G. Stevens

Mössbauer Effect Data Center, The University of North Carolina, Asheville, NC 28804-8511 United States

Determination of Einstein and Debye parametric temperatures from Mössbauer relative absorption is discussed. Experimental error limit of these parameters is calculated. Experimental requirement for possibility of model comparison is shown. Two-parameter development of an Einstein model is proposed.

1. Introduction

The Mössbauer effect is used to study the lattice dynamics through the temperature dependence of the recoilless fraction - f. Interpretation of crystal dynamical parameters determined from recoilless fraction needs to take into account zero-point vibrations and weighting of the phonon spectrum when f is calculated [1,2]. This makes Mössbauer parametric temperatures a nuclear resonant absorption point of view on a certain phonon spectrum model compared to calorimetric or X-ray diffraction approach [3]. Usability of basic models like the Debye approach is still discussed in the light of including effects of anharmonicity [4]. Abilities and limitations of the Mössbauer method to determine model parameters and model comparison need further exploration.

2. Relative Absorption

The following consideration is limited to thin sample experiments and harmonic models. The line shape for thin absorber is given as

$$N(v) = B\left[1 - F_s\frac{t_a}{2}\frac{(w/2)^2}{(v - v_o)^2 + (w/2)^2}\right]$$ (1)

where B - counts at infinite velocity,
F_s - recoilless fraction in the source,
t_a - effective thickness of the absorber,
v_o - absorption maximum position,
w - linewidth (full width at half-maximum).

The counts in the absorption maximum for $v = v_o$ in (1) is

$$N(v_o) = B\left[1 - F_s\frac{t_a}{2}\right].$$ (2)

The relative absorption is determined experimentally as

$$I_o = \frac{B - N(0)}{B}$$ (3)

And the error of I_o can be written, assuming $B >> 1$ and $N(v_o) \approx B$, is

$$\sigma(I_o) \cong \sqrt{\frac{2}{B}} \qquad\qquad (4)$$

From (2) and (3), it follows that

$$I_o = F_s \frac{t_a}{2} = F_s n \sigma_o f_a a \qquad\qquad (5)$$

where   n - number of element atoms per unit area,
    σ_o - Mössbauer resonant cross-section,
    a - isotopic abundance of resonant isotope,
    f_a - absorber recoilless fraction.

The determination of f_a from the experimental relative absorption is

$$f_a = \frac{I_o}{F_s n \sigma_o a}, \qquad\qquad (6)$$

and the theoretical recoilless fraction is

$$f_a = \exp\left\{-\frac{E^2_\gamma <x^2>}{(hc)^2}\right\} \qquad\qquad (7)$$

where E_γ - transition energy,
    $<x^2>$ - mean square displacement.

The determination of f_a from relative absorption involves knowledge of several factors and the precision of determination of the recoilless factor depends mainly on the relative errors of absorber thickness, recoilless fraction in the source and relative absorption in the spectrum. Out of these, usually only the relative absorption is under control in the experiment. This limits the possibility of the usage of recoilless factor itself for study of lattice dynamics. On the other hand the use of the ratio of recoilless factors through the ratio of relative absorptions at different temperatures leaves only one experimental error to control.

$$R = \frac{I_o(T_2)}{I_o(T_1)} = \frac{f_a(T_2)}{f_a(T_1)} \qquad\qquad (8)$$

Experimental error of R depends on relative absorption and square root of experimental statistics

$$\sigma(R) = \frac{2}{I_o \sqrt{B}} \quad (I_o(T_1) \approx I_o(T_2) = I_o) \qquad\qquad (9)$$

3. Einstein Model

The temperature dependence of the mean square displacement for the Einstein model [5] is given as

$$<x^2> = \frac{h^2}{km\Theta_E}\left[\frac{1}{2} + \frac{1}{e^{\Theta_E/T} - 1}\right] \qquad (10)$$

where k - Boltzman constant,
 h - Planck constant,
 m - nuclear mass.

The experimental value of R is a function of two experimental measurement temperatures and one parameter – Θ_E:

$$R = \exp\left\{-\frac{E^2\gamma}{kmc^2\Theta_E}\left[\frac{1}{e^{\Theta_E/T_2} - 1} - \frac{1}{e^{\Theta_E/T_1} - 1}\right]\right\} \qquad (11)$$

where T_1, T_2 - experimental measure temperatures.

It is possible to calculate numerically the reversed dependence of Θ_E and its error from (11)

$$\Theta_E = E\ (R, n, T_1, T_2) \qquad \sigma(\Theta_E) = Q_E\ (R, n, T_1, T_2)\ \sigma(R)$$

where   n - number of experimental measurements,
 T_1 - starting measurement temperature,
 T_2 - final measurement temperature.

These functions are plotted in Figure 1 for ^{57}Fe and ^{119}Sn.

By increasing the number of measurements at different temperatures within the same temperature range one can increase the accuracy for the determination of Θ_E. In the case of 5 measurements in the interval from 4 to 300K, one can obtain an accuracy in Θ_E of approximately 0.2K when $\Theta_E<60$K. In the case of 15 measurements in the interval from 4 to 300K, one can reach an accuracy in Θ_E of approximately 0.05K when $\Theta_E<30$K. This error jumps to 0.5K at 100K and 4K at 150K.

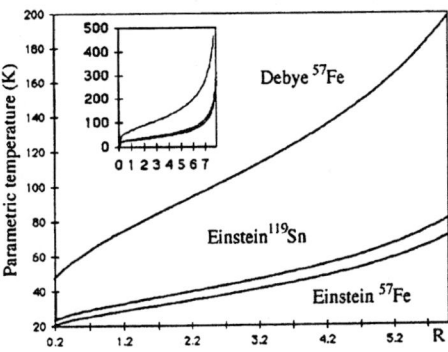

Fig. 1. Einstein temperatures for ^{57}Fe, ^{119}Sn and Debye temperature for ^{57}Fe vs R.

Calculations show that the same number of measurements made within different temperature regions extend the range of Θ_E that can be obtained with the same accuracy but not proportional to the increase in measurement temperature range. For example, when the experimental temperature region is increased by three times the maximum value of Θ_E is only increased by about 50%. The results are given in Figure 2.

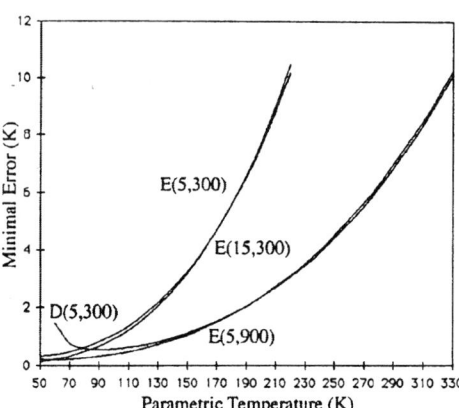

Fig. 2. Minimal error of model parametric temperature vs parametric temperature. E(5,300) corresponds to Einstein model applied to 5 measurements from 4 to 300K. E(15,300) - Einstein model for 15 measurements from 4 to 300K. E(5,900) – Einstein model for 5 measurements from 4 to 900K. D(5,300) – Debye model for 5 measurements from 4 to 300K. All graphs calculated assuming $I_o = 0.1$ and $B = 10^6$.

4. Debye Model

The temperature dependence of the mean square displacement for the Debye model [6] is given as

$$<x^2> = \frac{3h^2}{km\Theta_D}\left[\frac{1}{4}+\left(\frac{T}{\Theta_D}\right)^2 \int_0^{\Theta_D/T} \frac{xdx}{e^x-1}\right] \tag{12}$$

Using equations (7) and (8), R can be written as

$$R = \exp\left\{-\frac{3E^2_\gamma}{kmc^2\Theta_D}\left[\left(\frac{T_2}{\Theta_D}\right)^2 \int_0^{\Theta_D/T_2}\frac{xdx}{e^x-1} - \left(\frac{T_1}{\Theta_D}\right)^2\int_0^{\Theta_D/T_1}\frac{xdx}{e^x-1}\right]\right\} \tag{13}$$

R is a function of the experimental measured temperatures and the parameter – Θ_D. Numerical calculations allow reconstruction of the reversed dependence:

$$\Theta_D = D(R, n\, T_1, T_2) \qquad \sigma(\Theta_D) = Q_D(R, n, T_1, T_2)\, \sigma(R)$$

Analysis of these dependences leads to the following conclusions:

From 5 measurements in the interval of 4 - 300K the accuracy in Θ_D is 1K when $\Theta_D <$ 140K. When the number of measurements is 9 and the range is increased to 4 – 900K, one can estimate Θ_D with the same accuracy now up to 260K.

5. Comparison of Models

Comparison of these two models shows that in the same experimental conditions the Einstein model provides better accuracy, but in a more limited range of fitted model parameter than the Debye model. The use of Mössbauer spectroscopy for determining parameters of lattice dynamics models raise a question of distinguishability between models.

Einstein vs. Debye Model:

The Debye model predictions is fitted by adjusting Einstein model. The analysis gives the results, that there is a simple correlation between the two parameters. The existence of such correlation was discussed earlier [7]. Still there is a small unadjustable difference between two models expressed as follows:

$$X_{DE}(\Theta_D) = \sum_{i}^{n-1}(R_i^D - R_i^{E*})^2 \qquad (14)$$

    R^D - R values for the Debye model,
    R^{E*} - R values calculated from Einstein model fitted to Debye model.

This difference consists of intrinsic inequality depending on Θ_D and an experimental part which depends on the number of measurements and the measurement range. Lower measurement temperatures reveal additional difference. For example, five measurements from 77 to 300K give $X_{DE} = 4 \times 10^{-6}$ where as five measurements from 4 to 300K give $X_{DE} = 7 \times 10^{-4}$. More measurements do not increase X_{DE} but shift Θ_D of the maximum towards lower temperatures. Some of the above results are summarized in Figure 2.

Applying χ^2 criterion with 5% significance for 5 measurements gives χ^2 (4,0.05) = 9.5 requirement on error in R and minimal statistics for distinguishing between the two models can be obtained and is shown in Figure 3.

$$X_{DE}(\Theta_D) > \chi^2(5,0.95)\,\sigma^2(R)$$

$$B > B_{DE} = \frac{4\chi^2(5,0.95)}{X_{DE}(\Theta_D)I_o^2}$$

Estimation for ^{57}Fe gives the value of necessary counts of $B > 5 \times 10^6$ at $\Theta_D = 140K$

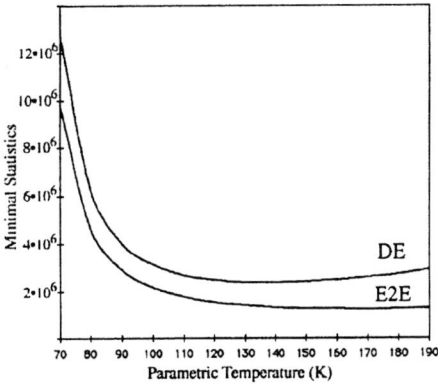

Fig.3. Minimal values of background statistics vs parametric temperature needed to distinguish between Einstein and Debye model - DE, and Einstein2 from Einstein model - E2E. In case of DE Debye temperature is a parametric argument.

Einstein2 vs. Einstein Model:

Einstein2 is a two parameter model which is probably the simplest step towards more complicated models of phonon spectra. It consists of two Einstein frequencies in the phonon spectrum. In calculation, one of the parameters is fixed at a certain value for Θ_1, while the other is changed from 50K to Θ_1. Then the single parameter Einstein model is fitted to it leaving nonadjustable discrepancy X_{E2E}. Analysis shows that wider temperature range reveal a greater difference between the models. A large number of measurements shift the position of the maximum X_{E2E} towards lower values of Θ_1 and does not increase the difference.

In case of ^{119}Sn Einstein2 Model with $\Theta_1 = 50$ and $\Theta_2 = 190$ it is fitted best by Einstein($\Theta_E = 130K$) still leaving difference of $X_{E2E} = 2 \times 10^{-3}$. Assuming 5% significance level and relative absorption $I_o = 0.1$ the requirement on statistics is $B > 8 \times 10^5$

In case of ^{57}Fe Einstein2(50,170) is fitted best by Einstein model with $\Theta_E = 119K$ leaving $X_{E2E} = 1.3 \times 10^{-3}$. With the same assumptions for the level of significance and relative absorption one has to collect $B > 4 \times 10^6$ for there to be a noticeable difference between the two models.

Einstein2 vs. Debye Model:

^{119}Sn Einstein2(5,190) is fitted best by Debye model with $\Theta_D = 228K$ leaving the difference of 2×10^{-3}. This translates into the requirement that $B > 3 \times 10^5$ as the theoretical limit of distinguishability between models.

References

[1] R.M. Housley and F. Hess, *Phys. Rev.* 146 (1966) 517.
[2] R.H. Nussbaum, in *Mössbauer Effect in Methodology*, ed. I.G.Gruverman, Plenum Press New York (1966) p. 3.
[3] T. Nakazawa, H. Inoue and T. Shirai, *Hyperfine Interac.* 55 (1990) 1145.
[4] K. Mahesh, *Phys. Stat. Sol.* 61 (1974) 695.
[5] G.K. Wertheim, *Mössbauer Effect,* Academic Press, New York (1964).
[6] ed. H. Frauenfelder, *The Mössbauer Effect: A Review-with a Collection of Reprints,* W.A.Benjamin, Inc., New York (1962).
[7] A.A. Bahgat, *Phys. Stat. Sol.* 63 (1981) K39.

MÖSSBAUER INVESTIGATION OF COMMERCIAL FLOAT GLASS USING SN^{119} MÖSSBAUER SPECTROSCOPY

J. G. Stevens[1], A. Khasanov[1] and A. M. Snider[2]

[1]*Mössbauer Effect Data Center, University of North Carolina, Asheville, NC 28804 United States*
[2]*PPG Industries, Inc., Harmarville, Pennsylvania 15238 United States*

Sn^{119} Mössbauer spectroscopy has been used to investigate a series of samples of commercial float glass in which the bulk concentration of tin is extremely low (bulk concentrations of <10 ppm). A variety of experimental procedures have been used to enhance the spectra including preparation of samples of various thicknesses, techniques in computer fitting, and other optimization procedures. Tin Mössbauer spectroscopy is an excellent technique for distinguishing Sn^{+2} and Sn^{+4} oxidation states present in these materials. With the improvements in obtaining spectra, investigations of these materials is now possible.

1. Introduction

Currently the most common industrial method for making float glass is known as the "float" process in which molten glass is poured onto molten Sn in a tank that is very long being of the order of 70 meters. The process is a continuous one as the molten glass moves along the long tank, it cools and is solidified upon reaching the other end of the tank. It is of interest to investigate the plate glass to gain insight into the migration of the small amounts of the tin into the bulk of the glass. Of particular interest is the oxidation states of the Sn in the glass matrix. This is important as the quality of the glass surface is related to the oxidation state of the tin. The concentration of the tin in the bulk of these samples is in the order of 10 ppm or less. The outer 5-10μ of surface that is in contact with the molten tin is enriched with the tin approaching surface concentrations of 1%. Spectrum from an initial sample of production glass is given in Figure 1. At the initial stage of the investigation, spectra of very poor quality were obtained (see Figure 1), but seen in the spectrum are several peaks that are barely observable. This investigation has led to several methodologies for enhancing the spectra of float glass in order to be able to obtain usable and useful data. The use of Mössbauer spectroscopy to investigate float glass has been reported before, but not for samples in which the concentration is as low as it is for the samples studied here [1].

2. Sample Preparations

The first sample that was investigated is a sample in which the side of the glass that is not exposed to the molten tin was removed in part by a careful grinding. More effort was put into this stage of the preparation of the samples to make the glass disks as thin as possible. Eventually samples having a thicknesses of about 0.07 mm were prepared and used. These very thin samples were investigated by stacking a multiple of two or more of the disks and then obtaining count rates for various sample thicknesses. From this data plot of a theoretical signal-to-noise ratio could be determined and is plotted in Figure 2. The result is that the optimum sample thickness is one containing about six glass disks giving a combined total thickness of 0.25 mm. Eventually samples were ground to less than 0.02 mm in thickness at which stage conventional handling the samples became impractical.

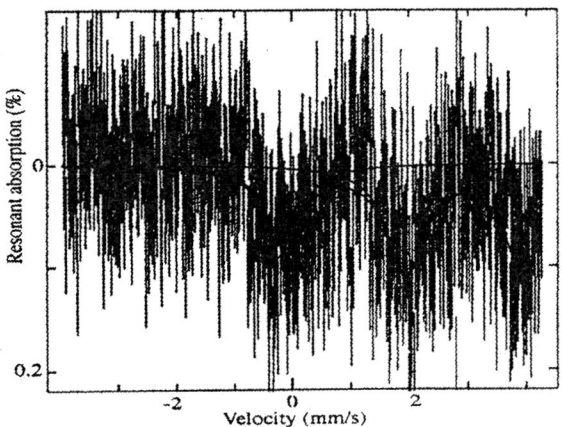

Fig. 1. Mössbauer spectrum of initial exploratory glass sample.

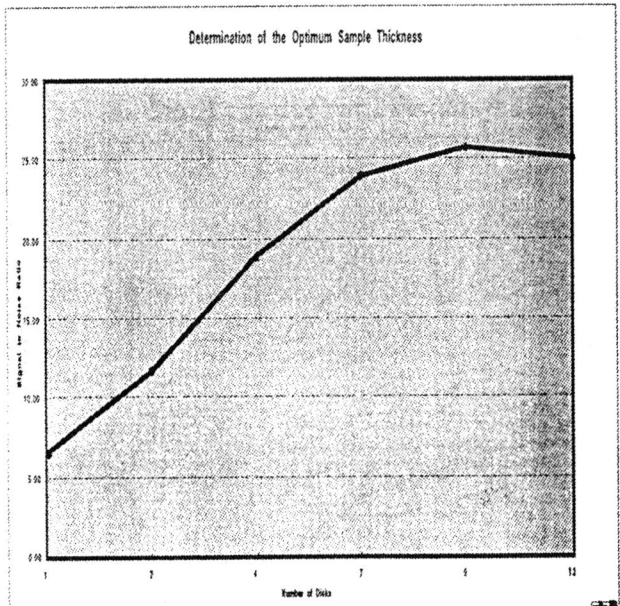

Fig. 2. Experimental data to determine the optimum sample thickness based on signal to noise considerations.

3. Spectra Fitting Procedures

One of the objectives with the investigation is being able to determine the relative amounts Sn^{+4} to Sn^{+2}. Even as the quality of the data improved with adjustments to sample thicknesses and the layering of an optimum number of disks, the resulting error in the determination of the area under the peaks was not of the desired values. It can be improved, if it is assumed that the Sn^{+2} and the Sn^{+4} sites in the various molten tin samples is essentially the same from one sample to another, one can then use the collection of data that was obtained on a series of several samples. This enables one to obtain the best values for several of the Mössbauer parameters (isomer shifts, widths, and quadrupole splittings). Then with these best values it is possible to fix these values in the parameter fittings. In general this

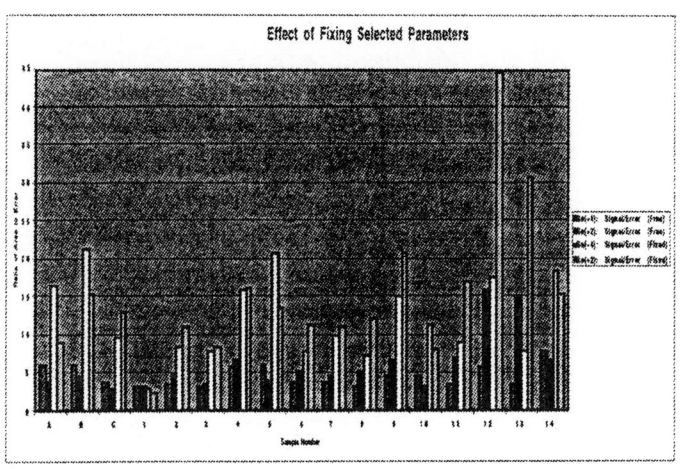

Fig. 3. The first two bars for each sample are the result of computer fitting with the Mössbauer parameters not constained. The second pair are the results of fixing certain Mössbauer parameters (see text). In general, there is an improvement by a factor of 2 to 3.

procedure provided a decrease in the relative errors in the range of a factor of two to three times. These results as shown in Figure 3. The basic assumption of there being only the one tin site has been examined. Through careful analysis of all of the data accumulated from the various glass samples investigated here, it is not possible to differentiate within the allowable statistics that there exists any other than one tin site.

Further correction in the determination of the relative amounts of Sn^{+2} to Sn^{+4} is that the Mössbauer fraction for these two sites is not the same even at liquid nitrogen temperatures. An assumption was made that the Mössbauer fraction of the Sn^{+2} and Sn^{+4} sites are very similar to the simple oxides of Sn. Using literature values for the Mössbauer fractions it is then possible to correct the determined relative amounts of Sn^{+2} to Sn^{+4} [2].

4. Source Intensity

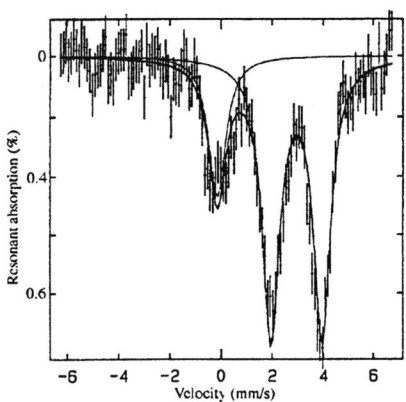

Fig. 4. Typical Mössbauer spectrum resulting from the methodologies described in the text.

Almost all the spectra were obtained with a source that had an intensity of less than 1 mC. A 10 mC source was most recently used, which provided considerably more counts. For example, previously counts per hour was somewhere in the range of between 4,000-14,000 counts per hour. The new source provides over 100,000 counts per hour. Going to more intense sources would not be possible as the count rate which the new source is providing is at the limits of the nuclear instrumentation. The resulting spectrum from all the improvements noted improvements in spectral quality it is possible to determine the relative amounts of the Sn^{+2} and Sn^{+4} oxidation states to within an error of about 1%.

References

[1] K.F.E. Williams, C.E. Johnson, J. Greengrass, B.P. Tilley, D. Gelder and J.A. Johnson, J. Non-Cryst. Solids 211 (1997) 164.
[2] J.A. Johnson, C.E. Johnson, K.F.E. Williams, D. Holland and M.M. Karim, Hyperfine Interact. 95 (1995) 41.

EVIDENCE OF RE-ENTRANT SPIN-GLASS IN $Fe_{0.5}Mn_{0.1}Al_{0.4}$ PHASE USING MOSSBAUER SPECTROSCOPY

Claudia González, J. A. Tabares, Ligia E. Zamora, A. García and G. A. Pérez Alcázar.

Departamento de Física Universidad del Valle. Cali, Colombia.
A. A. 25360 Fax 57 - 2 - 3393237

Mössbauer spectroscopy, ac susceptibility and ac calorimetry measurements carried out on the $Fe_{0.5}Mn_{0.1}Al_{0.4}$ disordered alloy, in the range of 12 to 300 K, showed two transition temperatures. One of them at about *30 K* linked to a re-entrant spin glass to ferromagnetic phase transition and other near to *215 K* associated to a ferromagnetic to paramagnetic phase transition.

1. Introduction

Systems presenting the spin-glass phase are characterized by disorder and also by competitive interactions of spins. These systems present also a critical temperature T_f at which the spins freeze randomly [1,2]. In the study of this phase some techniques are usually employed such as Mössbauer spectroscopy [3,4], ac magnetic susceptibility [3,5] and ac calorimetry [6].

It has been reported [7] that the freezing temperature T_f corresponds to the spin-glass to paramagnetic phase transition as the temperature is raised. Also, that the re - entrant spin-glass temperature T_k corresponds to the transition from the re-entrant spin-glass to ferromagnetic phase when T increases.

In the study of the spin-glass systems by Mössbauer spectroscopy, it has been observed that the mean hyperfine field (H) versus T curves behave in the same way as for the ferromagnetic systems [7,8]. This curve allows to obtain the critical temperature (T_f) at which $H = 0$. However, for re - entrant spin-glass systems, the H versus T curve exhibits a kink, which corresponds to the re - entrant spin-glass temperature T_k. Below this temperature the H values increase more rapidly [9] as the temperature is lowered.

Spin-glass and re-entrant spin-glass systems studied by ac magnetic susceptibility[3,5] reveal they are sensitive to the frequency and to the applied field.   This feature does not occur in ferromagnetic or antiferromagnetic systems.

The study of the $Fe_{0.5}Mn_{0.1}Al_{0.4}$ disordered alloy presented in this work is one of the *FeMnAl* alloys system in the bcc disordered phase. Previous experimental and theoretical studies of this system [10,11] have reported different magnetic phases, depending on composition and temperature.

2. Experimental procedure

The $Fe_{0.5}Mn_{0.1}Al_{0.4}$ alloy sample was prepared by melting pure (>99.9%) Fe, Mn, and Al powder in an arc furnace under argon atmosphere. After melting, the sample was evacuated sealed in a quartz tube and submitted to a heat treatment at 1000° C, during a week, in order to homogenize it. Finally, it was quenched in ice water in order to retain the high temperature phase. A powdered sample was obtained for Mössbauer spectroscopy measurements. Part of the sample was pressed to form a cylinder of 3 mm of diameter and 4 mm of height for magnetic susceptibility measurements. In order to carry out the calorimetric measurements a sheet of 1 mm^2 of area and 170 μm of height was prepared.

Mössbauer spectra were fitted, according to the disordered character of the sample, with hyperfine field distributions involving different isomer shifts. Using the Normos program with 45 sextets, adding a paramagnetic line carried out the fitting process.

3. Results and Discussion

Figure 1 shows the Mössbauer spectra (1a) taken at different temperatures. Spectra were fitted with hyperfine field distribution (1b) and a paramagnetic site.

From the fit of the spectra, the mean hyperfine field values were obtained for different temperatures. The H versus T curve is presented in figure 2. In this curve it can be observed a kink at $T_k \approx 30\,K$ and also that its extrapolation gives a $H = 0$ value for $T_C \approx 215\,K$. As it was previous discussed, this behavior is typical of a re - entrant spin-glass phase [4]. Thus, the extrapolated temperature T_C can be associated to a ferromagnetic to paramagnetic phase transition and the temperature for which the kink occurs, to a re-entrant spin-glass

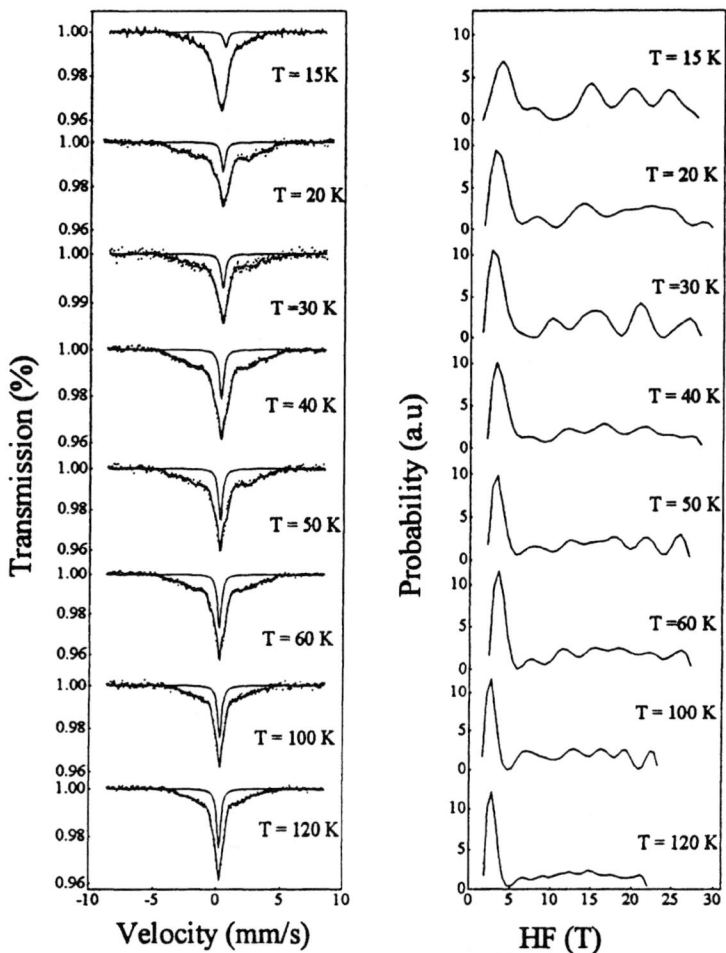

Fig. 1. Mössbauer spectra and hyperfine field distributions at several temperatures.

to ferromagnetic phase transition. The spin-glass behavior is possible due to the disorder as well as the competitive interaction of both ferromagnetic interactions (Fe-Fe) and antiferromagnetic interactions (due Mn). The additional increase of H values below T_k is attributed to the contribution of the freezed spins to the magnetization value [4, 5, 9].

Figure 3 shows the curves of ac magnetic susceptibility as a function of T. It can be observed that these curves present two peaks, near $T \approx 31\ K$ and $T \approx 203\ K$. These peaks are in agreement with the transition temperatures from re-entrant spin-glass to ferromagnetic and from ferromagnetic to paramagnetic phases, respectively, detected by Mössbauer spectroscopy. In the low temperature region, the peak positions are sensitive to frequency as is shown in the inset. This behavior is typical of a re-entrant spin -glass phase transition [12 - 14].

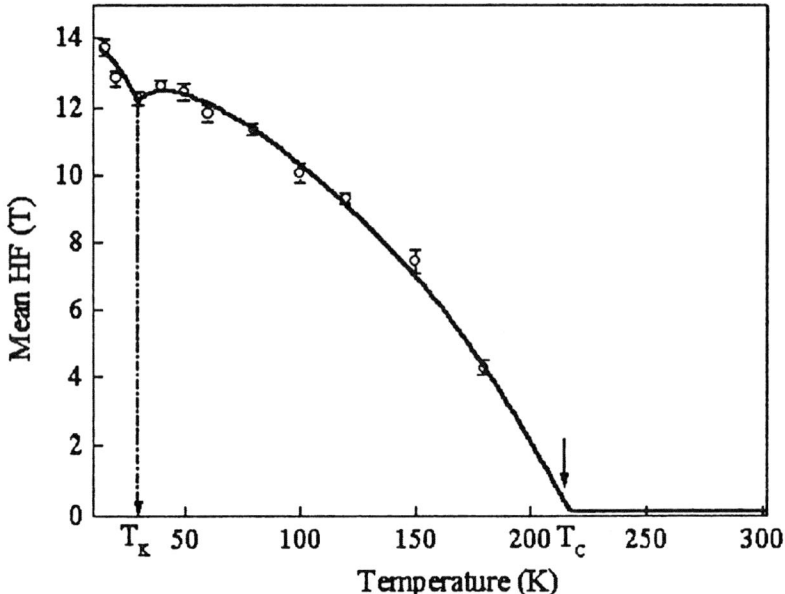

Fig 2. Mean Hyperfine Field H as a function of temperature.

Figure 4 shows the curves of specific heat as a function of T. In the range between *12* and *45 K*, no anomalies or changes of curvature are observable (fig. 4(a)). This is also in agreement with reported results in spin-glass systems by using specific heat measurement [6]. Those works reported that the specific heat versus T curves can not be used to detect T_k due that they do not present a well defined peak at this temperature. In figure 4(b) it can be observed these anomalies or maxima in $T \approx 216 K$ and $T \approx 223 K$, and these peaks are in the range of the transition temperature from ferromagnetic to paramagnetic phase detected by Mössbauer, and ac susceptibility measurements. So far it is not clear for us why in the ac calorimetric curve this two transition appears.

C. González et al. / Evidence of re-entrant spin glass in $Fe_{0.5}Mn_{0.1}Al_{0.4}$ phase using Mössbauer spectroscopy

81

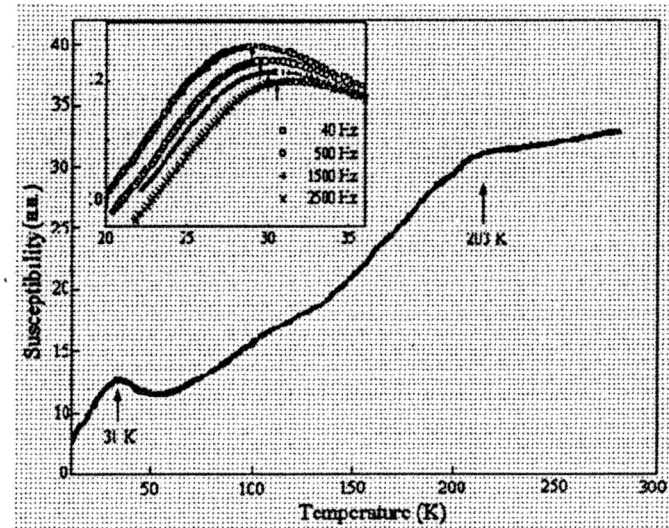

Fig.3. The ac susceptibility as a function of temperature in an ac field of 3 Oe and frequencies of 40, 500, 1500 and 2500 Hz.

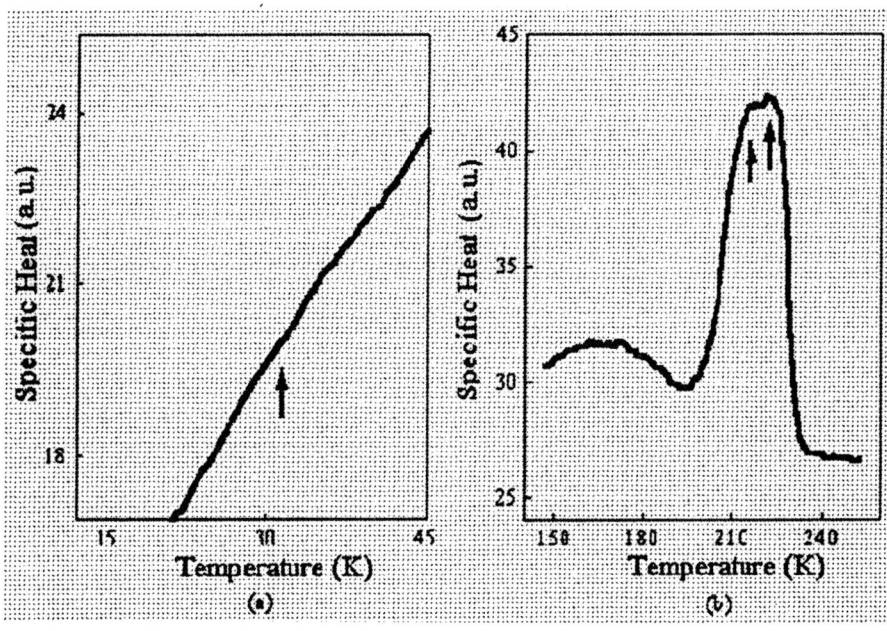

Fig. 4 Temperature dependence of the specific heat at (a) below 45 K and (b) above 140 K and below 260 K.

4. Conclusions

By using Mössbauer spectroscopy, ac susceptibility, and ac calorimetry techniques, it was possible to study the magnetic properties the $Fe_{0.5}Mn_{0.1}Al_{0.4}$ alloy sample. It was found that this sample presents two transition temperatures, one at $T \approx 30\,K$ which is associated to a transition from a re-entrant spin-glass phase to a ferromagnetic one, and the second at $T \approx 215\,K$ associated to a ferromagnetic to paramagnetic transition. The re - entrant spin behavior can be associated in this range of temperature, to the disorder as well as the competitive interaction of both ferromagnetic character (Fe - Fe) and antiferromagnetic character (Fe - Mn and Mn - Mn).

Acknowledgment

The authors would like to thank to COLCIENCIAS, Colombian Agency, and Universidad del Valle for their financial support.

References

[1]   K. Binder, and A. P. Young, Rev. Mod. Phys **58** (1986) 801.

[2]   K. H. Fischer, Phys. Stat. Sol. **B 116** (1983) 357.

[3]   J. Hesse, Hyp Int. **47** (1989) 357.

[4]   I. A. Campbell, Hyp Int. **27** (1986) 15.

[5]   C. A. M. Mulder, A. J. Van Duynveldt, and J. A. Mydosh, Phys Rev **B** (1981) 1384.

[6]   L. E Wender and P. H. Keesom, Phys Rev, B 13 (1976) 4053.

[7]   C. E Violet and Borg, R. L. Phys Rev **162** (1967) 608B.

[8]   F. Varret, A. Hamzic, and I. C. Campbell, Phys Rev **B 26** (1982) 5282.

[9]   D. H. Ryan, "Recent Progress in Random", Ed. D. H. Ryan (Word Scientics, Singapure 1992).

[10]   G. A. Pérez Alcázar, J. A. Plascak, and E. Galvão da Silva, Phys Rev **B 38** (1988) 2816.

[11]   Ligia E. Zamora, G. A. Pérez, A. Bohórquez, and J. A. Tabares, JMMM. **137** (1994) 339.

[12]   J. A. Mydosh, "Spin Glasses: an experimental introduction", Edit Taylor and Francis, London Washington, DC. (1993).

[13]   F. Borsa, M. G. Pini, A. Rettori, and V. Tognetti, Phys Rev **B 28** (1983) 5173.

[14]   Ch. Bötter, R. Starsch, A. Wulfes, and J. Hesse. JMMM 99 (1991) 280.

MÖSSBAUER EFFECT STUDY OF NITROGEN DISTRIBUTION IN Sm_2Fe_{17} AND Y_2Fe_{17} NITRIDES

C.E. Rodríguez Torres, F.H. Sánchez, M.B. Fernández van Raap and L.A. Mendoza Zélis

Departamento de Física, Fac. de C. Exactas, Universidad Nacional de La Plata, C.C. 67 (1900) La Plata, Argentina. email:torres@venus.fisica.unlp.edu.ar

Mössabuer effect studies of the $R_2Fe_{17}N_x$ with R = Y and Sm and $0 < x < 2.9$ have been performed at 85 K. The results are compared with previous NMR measurements. It was found that the existence of a nitrogenated/unnitrogenated configuration is consistent with the experiments. The agreement between Mössbauer and [89]Y spin-echo NMR results is good if it is assumed that at intermediate steps the nitrided region is ihnomogeneous and that the nitrogen fills the 6h and 12i sites simultaneously up to the 12i occupancy limit (1/6) and afterwards only the occupation of 6h sites increases.

I. Introduction

The compounds of R_2Fe_{17} (R=Y, rare earth) class have large magnetization values, but rather low ordering temperatures (T_c) and anisotropy constant. However, it has been shown that the absorption of N_2 by Sm_2Fe_{17} and Y_2Fe_{17} approximately doubles their T_c and changes the anisotropy of the former from planar to uniaxial[1]. In order to determine how the gas insertion produces this improvement, it becomes very important to study the systems on a microscopic scale.

$R_2Fe_{17}N_x$ compounds can be prepared with x values typically up to 2.9 retaining their original crystalline structure (rhombohedral Th_2Zn_{17} for Sm and hexagonal Th_2Ni_{17} for Y). Nitrogen accommodation and distribution are controversial matters. While some authors have reported the exclusive occupancy of 9e octahedral site[2,3] (9e in the rombohedral structuctured, 6h in the hexagonal one) other [4] suggested a partial occupation of a second one, the 18g (12i) thetrahedral site. Regarding the nitrogen distribution, it is an open question whether the metaestable nitride is a simple gas-solid solution with continuos range of intermediate nitrogen contents [5,6] or it is a two-phase mixture of nitrogen-poor and nitrogen-rich phases [7,8].

In the present paper we present the results of a [57]Fe Mössbauer effect (ME) study of the $Sm_2Fe_{17}N_x$ and $Y_2Fe_{17}N_x$ compositional series. A comparison between these and previously published [89]Y spin-echo NMR results is performed[8].

II. Experiment

All the studied samples were supplied by Dr Y.D. Zhang from the Department of Physics of the Connecticut University. The ingot alloys were made by arc melting. The nitrogenations were performed by annealing samples with different particle sizes at 460 °C for 17 h (for $Y_2Fe_{17}N_x$) and at 430 °C for 15 h (for $Sm_2Fe_{17}N_x$) under an ultrahigh purity N_2 (99.999%) gas flow at a pressure of 1 bar. These resulted in nominal N content values of x = 0.6, 1.2, 1.8, 2.44, and 2.8 for Y and x = 0.28, 0.66, 1.31, 1.79, 2.26 and 2.66 for Sm.

The [57]Fe Mössbauer effect measurements were performed at 85K under transmission geometry with a standard constant acceleration spectrometer having a [57]Co*Rh* radioactive source.

III. Results and discussion

Fig. 1. shows the ME spectra for $Sm_2Fe_{17}N_x$ and $Y_2Fe_{17}N_x$ series. The Sm_2Fe_{17} spectrum was fitted

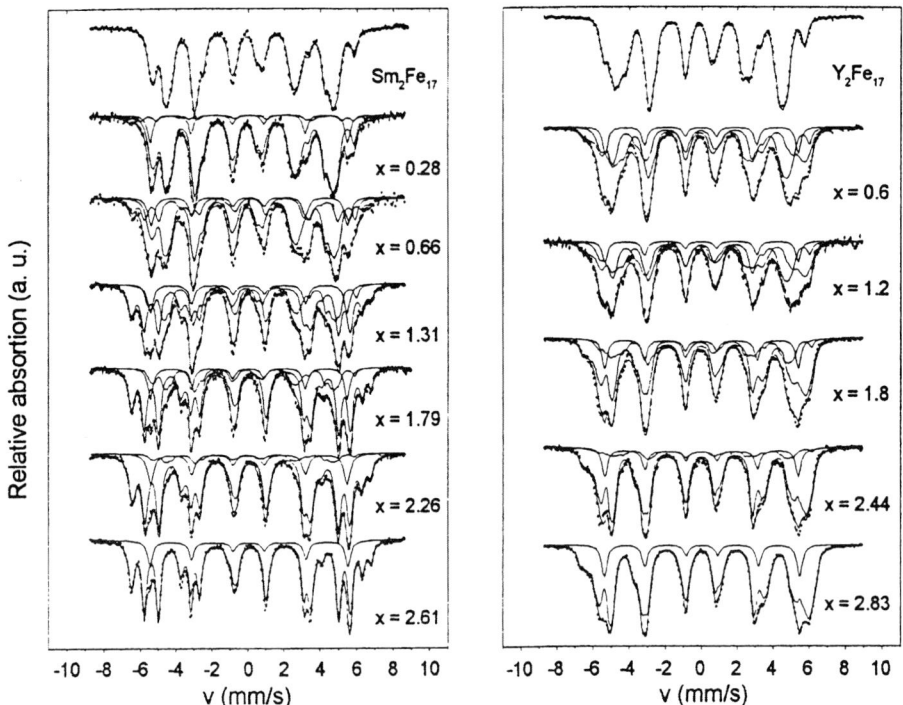

Fig. 1 Mössbauer effect spectra of Y₂Fe₁-, Y₂Fe₁₇Nₓ (right) and Sm₂Fe₁₇, Sm₂Fe₁₇Nₓ (left) at 85 K. The three theoretical subspectra correspond to the R₂Fe₁₇ hidrogenated and unnitrogenated regions and to α-Fe

with seven independent sextets: one of them corresponding to the dumbbell site (6c) and the other to the 9d, 18h and 18f, each one splitted into two groups with an intensity ratio 2:1 due to a combined effect of the dipolar field and the quadrupole interaction. In the case of the $Sm_2Fe_{17}N_{2.61}$ (fully nitrogenated) the magnetization direction is along the c-axis, then the angle between the hyperfine field and the electric field gradient is the same for all crystallographic equivalent sites. There is no additional splitting and the spectrum could be fitted with only four interactions corresponding to the four inequivalent sites of the 2:17 structure and an additional one associated with iron in α-phase. The presence of α-Fe in the nitrogenated phase will be discussed below.

In the spectra of Y_2Fe_{17} and $Y_2Fe_{17}N_{2.83}$, the 4f site was fitted with one sextet, whereas each of the three remaining sites were fitted with three sextets of equal area. This analysis is similar to that used by F. Grandgean and G. Long for $Th_2Fe_{17}N_{2.6}$ [9] and for Y_2Fe_{17} and $Y_2Fe_{17}N_{2.6}$[10]. All attempts to fit these spectra with fewer than ten interactions were unsuccessful. The fit with ten sextets indicates that the magnetization lies in a general direction of the hexagonal basal plane of these compounds. In the case of the nitride samples there was another interaction associated with iron in α-Fe.

The resulting parameters are shown in Table I and are in good agreement with physical considerations. The justification of the assignment of hyperfine parameters to structural sites as well as hyperfine parameter changes with volume expansion and nitrogen hybridization effects is discussed elsewhere[11].

TABLE I

⁵⁷Fe hyperfine fields (averaged over subspectra) and isomer shifts (relative to α-Fe) at each crystallographic site for the R₂Fe₁₇ and the fully nitrogenated compound

Compound	Hyperfine parameter *	Site			
		6g(9d)	12k(18h)	12j(18f)	4f(6c)
Y₂Fe₁₇	B_{hf} (T)	31.9	26.5	28.9	33.7
	δ_{IS} (mm/s)	-0.127	0.007	0.018	0.324
Y₂Fe₁₇N₂.₆₁	B_{hf} (T)	39.5	35.4	31.4	36.4
	δ_{IS} (mm/s)	-0.028	0.149	0.099	0.278
Sm₂Fe₁₇	B_{hf} (T)	30.3	26.4	30	35.4
	δ_{IS} (mm/s)	-0.064	0.23	0.055	0.027
Sm₂Fe₁₇N₂.₈₃	B_{hf} (T)	39.5	31.1	35.5	41.6
	δ_{IS} (mm/s)	0.02	31.1	35.5	0.27

* Typical errors for hyperfine fields and isomer shift are ≈ 0.1 T and ≈ 0.001 mm/s, respectively

Taking into account the previously obtained ⁸⁹Y spin-echo NMR results[8], where a nitrided/unnitrided (N/UN) region configuration was proposed, being the absorbed nitrogen atoms located in the nitrided regions of a sample particle, thus leaving the remaining part of the particle completely unnitrided, we attempted to fit the spectra as a mixture of R₂Fe₁₇ and R₂Fe₁₇Nᵧ and α-Fe. Then a fitting model should be proposed including: six or ten sextets for the unnitride phase, four or ten sextets for the fully nitride and another one for α-Fe. Such a large number of fitting parameters being impracticable, we attempted to fit these spectra with a superposition of three components: a simulated spectrum of the ingot alloy (x=0), a simulated spectrum of the fully nitrogenated sample (x=2.61 and 2.81 for Sm and Y respectively) and a sextet associated with iron in α-phase. The fitted parameters were: the relative fraction of the three components, the average isomer shift and hyperfine field of the simulated components and the hyperfine parameters of the sextet (isomer shift δ_{IS}, quadrupole shift ε, line width Γ). With this procedure the spectra could be rather well reproduced, better those of the Sm₂Fe₁₇Nₓ series than those of Y₂Fe₁₇Nₓ. The amount of ⁵⁷Fe in the N/UN phases was determined assuming equal recoil free fractions for all iron probes. In fig. 2 the evolution of this fraction, obtained from the ME analyses and from the equation

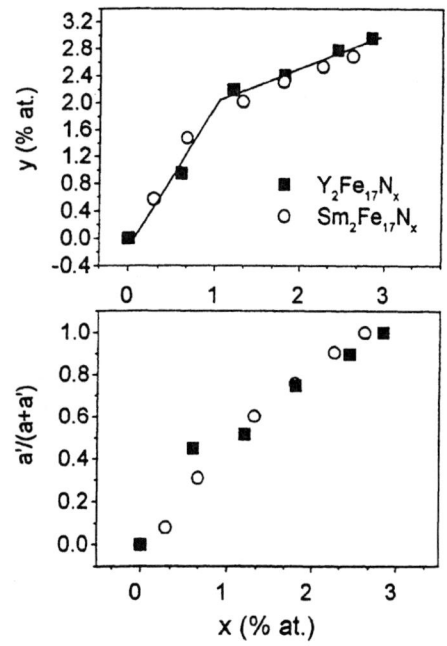

Fig. 2 Average nitrogen concentration and relative fraction of the nitrogenated region.

$$R_2Fe_{17} + \frac{x}{2}N_2 \Rightarrow a\,R_2Fe_{17} + a'\,R_2Fe_{17}N_y + b\,RN + c\,\alpha - Fe,$$

is shown, being y the average nitrogen concentration in the N region. The apparition of metallic iron is due to the partial disproportionation R₂Fe₁₇ + N into RN and α-Fe. Fig. 3 shows the hyperfine parameters obtained from the fits.

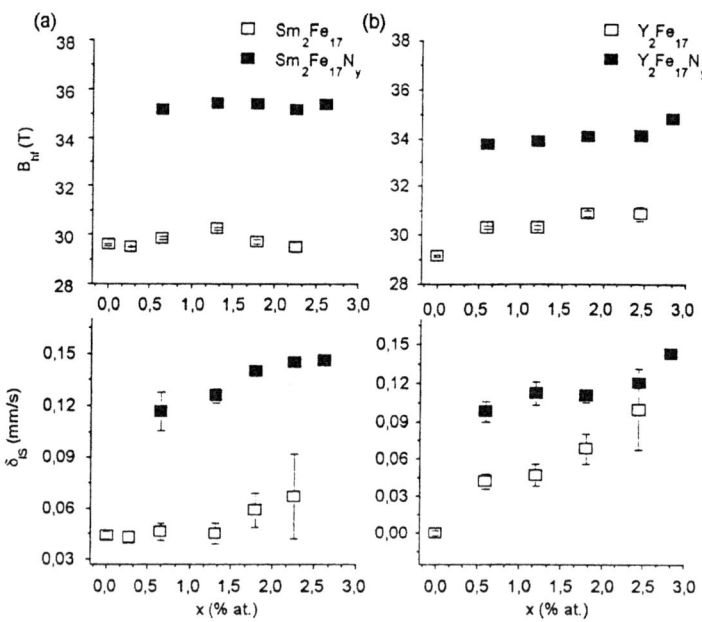

Fig. 3 Average hyperfine fields and isomer shifts obtained from the fits of the spectra shown in Fig.1, for the R_2Fe_{17} nitrogenated and unnitrogenated regions. (a) R=Sm, (b) R=Y

The ME results are consistent with a N/UN configuration. Already for the first nitrogenation steps the average hyperfine field (B_{hf}) of the interaction corresponding to iron in the N region is nearly equal to that obtained when the nitrogenation is completed. This indicates that either the N region has already its final nitrogen/Fe ratio or that at least this region has experienced a volume expansion approximately equal to that of the fully nitrogenated one. We assume that the second hypothesis holds due to the evolution of the B_{hf} with y (see fig. 3). NMR results on Y indicate that sites with 0 and 2 nitrogen atoms near neighbors (nn) are almost exclusively present[8], whereas our [57]Fe ME measurements indicate that $0 < y < 2$ in the N region for $0 < x < 1.2$. Therefore the N region must be inhomogeneous in the sense that it must host R sites with both 0 and 2 nitrogen nn. Considering that in the 0 nitrogen zones (within the N region) the volume expansion has already taken place (as in the 2 nitrogen ones), the [57]Fe within the 0 nitrogen region will experience a B_{hf} similar to that of those in the 2 one. Bitter pattern observations on $Sm_2Fe_{17}N_x$ indicates that nitrogen diffusion is anisotropic giving rise to a parallel lines type domain pattern with deeper penetration in the c axis direction[12]. This produces a heterogeneous nitrogen distribution in the c plane, where zones with 0 and 2 nitrogen nn to a R atom alternate.

In order to compare the ME results with those obtained by NMR in the $Y_2Fe_{17}N_x$ series, we estimate Y(0), Y(1), Y(2) and Y(3), where Y (Z) is the fraction of Y atoms with Z nitrogen nn (in a 6h interstitial site) from the relative fractions of iron in N and in UN regions and the estimated average nitrogen concentration for the N region. Following the NMR results[8] we assume that Y(1) = 0 in all cases. Taking into account the discussion above, within the N region two set of assumptions can be made according to the value of y.

For $y < 2$: (i) there are uY with 2 nitrogen nn (Y(2)) and vY with 0 (Y(0))

(ii)Y(1) and Y(3) are approximately equal to zero.

Therefore, the average nitrogen concentration obtained from ME results will be:

$$y = \frac{u \times 2 + v \times 0}{u + v} \qquad\qquad u + v = \frac{a'}{a + a'}$$

For y > 2: (i) Y(2) = u and Y(3) = w

      (ii) Y(0) and Y(1) are approximately equal to zero

Hence,

$$y = \frac{u \times 2 + w \times 3}{u + w} \qquad\qquad u + w = \frac{a'}{a + a'}$$

Additionally we must decide how much nitrogen is in the 6h site and how much in the 12i. All the studies indicate that nitrogen predominantly occupies the 6h octahedral sites. Whereas some authors report the exclusive occupancy of these sites[3] other suggest the partial occupation of a second interstitial site, the 12i. Jaswall et al [4] have reported that the nitrogen fills the 12i site first, to its occupancy limit 1/6 and then fills the 6h site to a limit of 2/3. A different model was given by Yang et al[13] who concluded that nitrogen filled the 12i and 6h sites simultaneously and achieved a final occupancy of 0.35 and 0.8, respectively. On the other hand, Zhang et al have estimated (from degassing experiments) that approximately between 4-7 % of the total nitrogen is in 12i sites. To calculate the histograms for Y(Z) we propose different models which take into account these hypothesis:

Model 1: nitrogen fills the 6h and 12i sites simultaneously being the fraction of nitrogen in the 12i about a 7% of the total.

Model 2: nitrogen fills the 6h and 12i sites simultaneously being the fraction of nitrogen in the 12i about a 33% of the total.

Model 3: nitrogen fills the 6h and 12i sites simultaneously up to the 12i occupancy limit (1/6) and afterwards only the occupancy of 6h sites increases.

Model 4: nitrogen fills the 6h sites first to its occupancy limit (2/3) and then the 12i are occupied.

The last two models are supported by the results shown in fig. 2 where it can be seen that the average nitrogen concentration y has a kink for y $\cong$ 2: increases rapidly up to y $\cong$ 2 and then

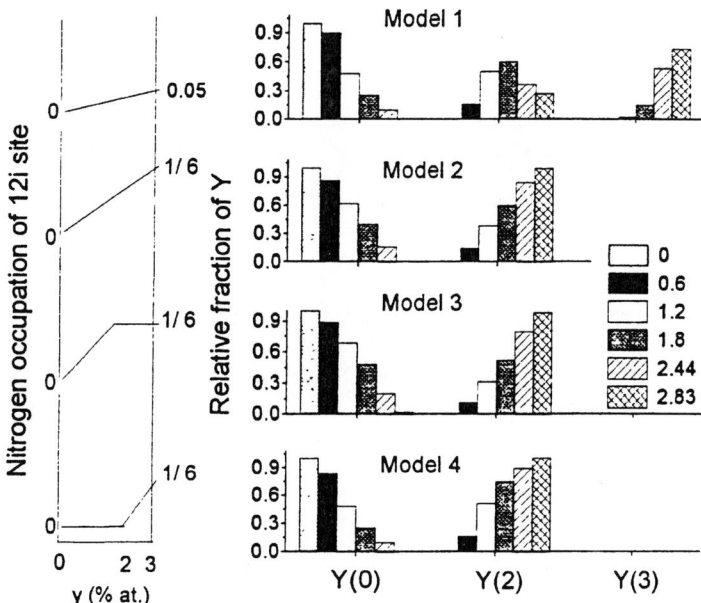

Fig. 4 Histograms of relative fraction of Y with 0,2, and 3 nitrogen nn at 6h sites. In the left a schematic representation of the models for the 12i site nitrogen occupation is shown.

increases slowly up to y $\cong$ 3 (where the nitrogenation is completed).
The resulting histograms are shown in fig. 4. Within the first model, for the completely nitrogenated sample, 65 % of yttrium would have 3 nitrogen nn. This is contrary to what is observed by NMR, where only the signal coming from Y(2) appears. In the other three models the experimentally obtained condition Y(3) $\cong$ 0 is satisfied, but the relative ratio Y(2):Y(0) varies from one to another especially for intermediate nitrogen concentrations. In this regard the model which reproduces better the NMR results is the third one. For instance, for x = 1.8 the ratio Y(2):Y(0) takes the values 1.5, 1.0 and 4 for models 2, 3 and 4, respectively, while the experimental NMR value is 0.81.

IV.   Conclusions

In the two cases (Y and Sm) the ME results support the hypothesis of the existence of a UN/N configuration for partially nitrogenated states. The ME and NMR results are in agreement under the assumption that the N regions are inhomogeneous at intermediate nitrogen concentrations and the nitrogen fills the 6h and 12i sites simultaneously up to the 12i occupancy limit (1/6) and afterwards only the occupation of 6h sites increases.

Acknowledgments

The authors gratefully acknowledge the partial financial support from CONICET (through Programa TENAES) of Republica Argentina and from the Secretariat of Science of the European Community.

[1] J.M.D. Coey and H. Sun, *J. Magn. Magn. Mater*, **87** (1990) L251.
[2] R.M. Ibberson, O. Moze, T.H. Jacobs and K.H.J. Buschow, *J. Phys.: Condens. Matter* **3** (1990) 1219.
[3] O. Isnard, S. Miraglia, J.L. Soubeyroux, D. Frunchard and J. Pannetier, *Phys. Rev. B* **45** (1992) 2920.
[4] S.S. Jaswal, W.B. Yelon, G.C. Hadjipanayis, Y.Z. Wang and D.J. Sellmyer, *Phys. Rev. Lett.* **67**(5) (1991) 644.
[5] R. Skomski and J.M.D. *Coey, J. Appl. Phys.,* **73** (1993) 7602.
[6] C.C. Colucci, S. Gama, C.A. Ribeiro and L.P. Cardoso, *J. Appl. Phys* **75** (1994) 6003.
[7] J.M.D. Coey, R. Skomski and S. Wirth, *IEEE Trans. Mag.,* **28** (1992) 2332.
[8] Y.D. Zhang, J.I. Budnick, D.P. Yang, G.W. Fernando, W.A. Hines, T.D. Xiao and T. Manzur, *Phys. Rev. B* **51** (1995) 12 091.
[9] G.J. Long, O.A. Pringle, F. Grandjean, W.B. Yelon and K.H.J. Buschow *J. Appl. Phys* **75** (1994) 2598.
[10] F. Grandjean, G.J. Long, O.A. Pringle and K.H.J. Buschow, *Hyperfine Interacttions* **94** (1994) 1971.
[11] C. E. Rodríguez Torres, F.H. Sánchez, M.B. Fernández van Raap and L.A Mendoza Zélis, unpublished
[12] C. Christodoulou and N Komada, *J. Appl. Phys.* **76** (1994) 6041.
[13] Q.W. Yan, P.L. Zhang, Y.N. Wei, K. Sun, B.P.Hu, Y.Z. Wang, G C. Liu, C. Gau and Y.F. Chen, *Phys. Rev. B* **48** (1993) 2878.

HYPERFINE INTERACTION OF ^{57}Fe AND AVERAGE Fe MAGNETIC MOMENTS IN Tb$_3$Fe$_{29-x}$Cr$_X$ AND Y$_3$Fe$_{29-x}$T$_X$ (T=V AND Cr)

Xiu-Feng Han[1,3], L.C.C.M. Nagamine[1], E. Baggio-Saitovtch[1], I. Souza Azevedo[1], Hong-Li[2], Yi-Wei Zheng[2], and Lan-Ying Lin[3]

[1]*Centro Brasileiro de Pesquisas Físicas, Rua Dr. Xavier Sigaud 150, Urca, 22290-180, Rio de Janeiro, RJ, Brazil*
[2]*Department of Physics, Beijing Normal University, Beijing 100875, P. R. China*
[3]*Institute of Semiconductors, Chinese Academy of Sciences, Beijing 100083, P. R. China*

^{57}Fe Mössbauer spectra in Tb$_3$Fe$_{29-x}$Cr$_x$ (x=1.0, 1.5, 2.0 and 3.0) and Y$_3$Fe$_{29-x}$T$_x$ (T=V and Cr) were collected for several temperatures from 4.2K up to 300K. The spectra analysis was based on previous results of magnetization and neutron powder diffraction measurements. The average Fe magnetic moments at 4.2 K deduced from our data are in accord with magnetization measurement. The ratio of 14.5 and 13.8 T/μ_B between the average hyperfine field and the iron magnetic moment, obtained from our data in Y$_3$Fe$_{29-x}$V$_x$ and Y$_3$Fe$_{29-x}$Cr$_x$, are in agreement with a ratio of 14.5 T/μ_B deduced from the R$_n$T$_m$ alloys by Gubbens et al. The average hyperfine field of Tb$_3$Fe$_{29-x}$Cr$_x$ (x=1.0, 1.5, 2.0 and 3.0) at 4.2 K and 300K decreased with Cr concentration, which is also in accordance with the variation of the average Fe magnetic moment in the Tb$_3$Fe$_{29-x}$Cr$_x$ compounds.

1. Introduction

In the last years, many experimental studies on the interstitial nitrides[1,2], carbides[2-4] and hydrides[4,5] of the R$_3$Fe$_{29-x}$T$_x$ compounds (R=rare earth; T=transition metal) have been performed in order to develop new rare earth permanent magnet materials. The nitrides Sm$_3$(Fe,T)$_{29}$N$_y$ (T=Ti, V, Cr and Mo) and the carbides Sm$_3$(Fe,T)$_{29}$C$_y$ (T=Ti, V, Cr and Mo) were regarded as new candidates for hard magnet application, which further inhanced the interest in the structure and magnetic properties of R$_3$Fe$_{29-x}$T$_x$ compounds. In this work, we report the ^{57}Fe Mössbauer spectroscopy measurements on Tb$_3$Fe$_{29-x}$Cr$_x$ (4.2K, 77K, 150K, 200K and 300K) and on Y$_3$Fe$_{29-x}$T$_x$ (T=V and Cr) at 4.2K and 300K, and the deduced values for the average Fe magnetic moment at each temperature.

2. Experimental Methods

Polycrystalline samples of Tb$_3$Fe$_{29-x}$Cr$_x$ (x=1.0, 1.5, 2.0 and 3.0) and Y$_3$Fe$_{29-x}$T$_x$ (T=V and Cr) were prepared and characterized as described in our previous work[6,7]. The ^{57}Fe Mössbauer spectra were collected at room temperature (RT) and at 4.2K with the sample and ^{57}Co/Rh γ-ray source at the same temperature respectively. The spectra were analyzed with only five sub-spectra, based on the results of x-ray diffraction[7] (XRD), neutron powder diffraction[8] (NPD) and magnetization measurements[7]. The value of saturation magnetization for the Tb$_3$Fe$_{28.0}$Cr$_{1.0}$ compound at RT was corrected to 72.0 Am^2/kg (27.0μ/f.u.), derived from M versus 1/H plots by extrapolating 1/H to zero, based on the high-field data of the magnetization curves for the magnetically aligned samples measured in pulsed magnetic fields (PMF)[6]. This value is in accordance with the result of the magnetization for the free powder samples, measured recently with a Superconducting Quantum Interference Device (SQUID) magnetometer.

3. Results and Discussion

The crystalline structure of R$_3$Fe$_{29-x}$T$_x$ compounds has 11 sites for the transition element M. The Fe ions occupying the 11 sites can be divided into five groups, according to the Fe- and R-atom numbers of nearest neighbors (NN) and the average bond length. These constraints were imposed to reduce the number of fitting parameters. Relative intensity was constrained to 3:2:1 ratio and the initial area ratio is shown in

Table 1. The values were based on the NPD results[8] for Nd$_3$Fe$_{24.52}$Cr$_{5.48}$ in the case of Tb compounds. The Mössbauer spectra with the fits are shown in Figure1 (Y$_3$Fe$_{27.2}$Cr$_{1.8}$) and Figure 2 (Tb$_3$Fe$_{29-x}$Cr$_x$). The hyperfine fields of each sub-spectra, the average hyperfine fields and the deduced Fe magnetic moments are shown in Table 2.

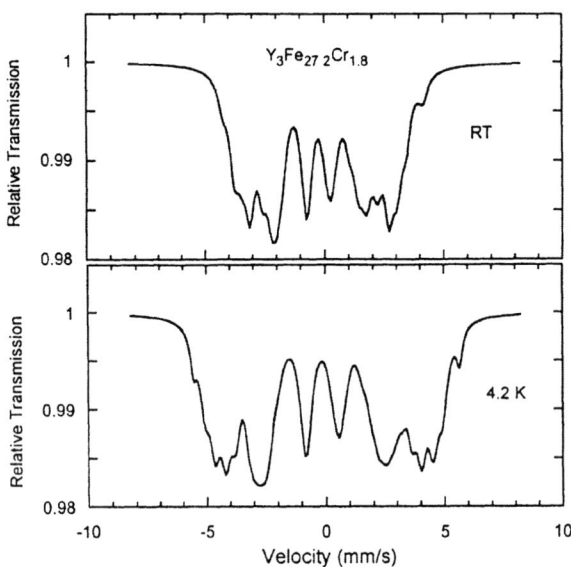

Fig 1. Mössbauer spectra of Y$_3$Fe$_{27.2}$Cr$_{1.8}$ at RT and 4.2K.

Table 1. The five groups of "Fe-types", their Fe occupancy number (FON), and the initial area ratio (A) used for the Mössbauer fits of the Nd$_3$Fe$_{23.52}$Cr$_{5.48}$ and Tb$_3$Fe$_{26.0}$Cr$_{3.0}$ compounds.

Nd$_3$Fe$_{23.52}$Cr$_{5.48}$	FON	A (%)	Tb$_3$Fe$_{26.0}$Cr$_{3.0}$	FON	A (%)
Fe$_3$, Fe$_6$, Fe$_7$	5.34	11.4	Fe$_3$, Fe$_6$, Fe$_7$	8.36	16.1
Fe$_5$, Fe$_{10}$	12.00	25.5	Fe$_5$, Fe$_{10}$	11.00	23.10
Fe$_8$, Fe$_9$	13.14	27.9	Fe$_8$, Fe$_9$	14.43	27.7
Fe$_1$, Fe$_{11}$	9.60	20.4	Fe$_1$, Fe$_{11}$	9.78	18.8
Fe$_2$, Fe$_4$	6.97	14.8	Fe$_2$, Fe$_4$	7.44	14.3

Table 2. Average hyperfine fields (B$_{hf}$) from Mössbauer measurements at 4.2K for Y$_3$Fe$_{27.2}$Cr$_{1.8}$ and Tb$_3$Fe$_{29-x}$Cr$_x$ (x=1.0; 1.5; 2.0; and 3.0), relative to 5 "Fe-type" sites.

R$_3$Fe$_{29-x}$Cr$_x$	B$_{hf}$(1) (T)	B$_{hf}$(2) (T)	B$_{hf}$(3) (T)	B$_{hf}$(4) (T)	B$_{hf}$(5) (T)	B$_{hf}$ (T)	μ_{Fe}^{MS*} (μ_B/Fe)	μ_{Fe}^{MS**} (μ_B/Fe)	μ_{Fe}^{M} (μ_B/Fe)
Y$_3$Fe$_{27.2}$Cr$_{1.8}$	34.1	29.9	26.8	23.8	19.6	26.5	1.83	1.92	1.92
Tb$_3$Fe$_{28.0}$Cr$_{1.0}$	38.5	33.6	30.7	27.8	25.0	30.0	2.07	2.17	1.90
Tb$_3$Fe$_{27.5}$Cr$_{1.5}$	35.0	31.6	29.0	26.1	22.7	28.2	1.94	2.04	1.80
Tb$_3$Fe$_{27.0}$Cr$_{2.0}$	34.6	30.4	26.8	23.8	19.8	26.7	1.84	1.93	1.79
Tb$_3$Fe$_{26.0}$Cr$_{3.0}$	31.6	26.2	26.5	22.1	18.5	24.2	1.67	1.75	1.64

We have roughly, $B_{hf}(1) > B_{hf}(2) > B_{hf}(3) > B_{hf}(4) > B_{hf}(5)$, which seems to be in agreement with the number of Fe and R nearest neighbors atoms at each of those Fe sites, except for $B_{hf}(2)$. $B_{hf}(1)$ corresponds to the Fe dumbbell sites with the largest Fe-atom NN numbers (13) and the largest Fe magnetic moment. $B_{hf}(2)$ corresponds to the Fe$_5$ and Fe$_{10}$ sites with the smallest Fe-atom NN numbers (9), but with the largest R-atom NN numbers (3) which could indicate a significant transferred spin polarization induced by the neighboring rare earth.

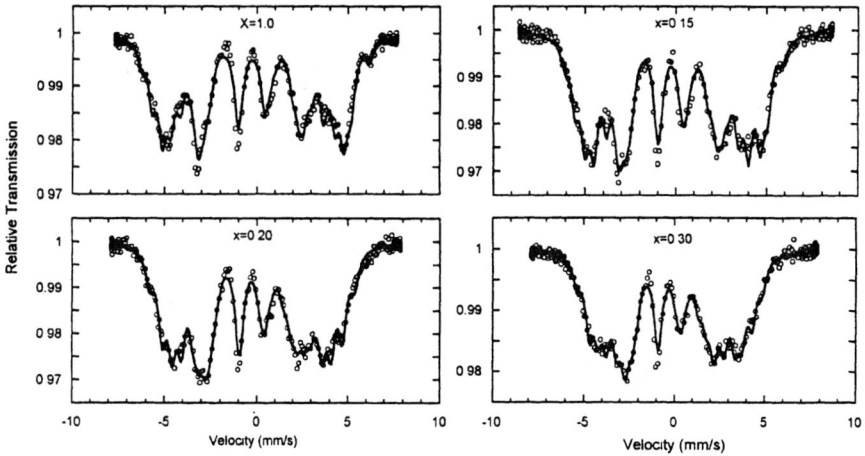

Fig 2. Mössbauer spectra of Tb$_3$Fe$_{29-x}$Cr$_x$(x=1.0, 1.5, 2.0, 3.0) at 4.2 K

Assuming a ratio of 14.5 T/μ_B between B_{hf} and μ_{Fe} deduced from our results for Y$_3$Fe$_{27.4}$V$_{1.6}$ and also from Mössbauer studies in the R$_x$Fe$_y$ alloys[9], the average magnetic moments (μ_{Fe}^{MS*}) of R$_3$Fe$_{29-x}$Cr$_x$ can be estimated and are shown in Table 2.

The average Fe magnetic moments as deduced from Mössbauer and magnetization measurements are in very good agreement. However, if we use a ratio of 13.8 T/μ_B deduced from our Mössbauer results for Y$_3$Fe$_{27.2}$Cr$_{1.8}$, the average Fe magnetic moments $\left(\mu_{Fe}^{MS**}\right)$ are slightly larger than the results of the magnetization measurements. Therefore, the average Fe magnetic moments decrease linearly with increased the Cr concentration, supporting the suggestion given by Han et al.[6] that the 3:29 structure is an ordered combination of the 2:17 and 1:12 structures in a ratio 1:1. This can be expressed as:

$$\mu_{Fe}(3) = \mu_{Fe}(2) - \frac{x\left[\mu_{Fe}(2) - \mu_{Fe}(1)\right]}{X_{max}}$$

where $\mu_{Fe}(3)$, $\mu_{Fe}(2)$ and $\mu_{Fe}(1)$ are the average Fe moments in 3:29, 2:17 and 1:12 compounds, and X_{max} is the highest Cr concentration which was used.

This empirical formula seems interesting, and it indicates that there exists a correlation among the intrinsic magnetic properties of the R$_3$Fe$_{29-x}$T$_x$, R$_2$Fe$_{17}$, and RFe$_{12-x}$T$_x$ structures. Therefore, a higher average magnetic moment, i.e., a higher saturation magnetization can be achieved for the R$_2$Fe$_{17}$ instead of RFe$_{12-x}$Cr$_x$ compounds. This may be due to the variation of spin up and spin down Fe-3d density of states at the Fermi level, E_F, associated with the mixture of the Cr valence electrons. The increasing Cr concentration in the 3:29, 2:17, and 1:12 compounds leads to a decrease of the average Fe magnetic moments.

The calculated values of average Fe hyperfine fields and the corresponding magnetic moments for $Y_3Fe_{27.2}Cr_{1.8}$ and $Tb_3Fe_{29-x}Cr_x$ (x=1.0, 1.5, 2.0 and 3.0) at 300 K are shown in Table 3. We assumed a ratio of 13.45 T/μ_B between B_{hf} and μ_{Fe}, obtained from the Mössbauer fitting for $Y_3Fe_{27.2}Cr_{1.8}$ compound at 300 K. The average Tb magnetic moments, μ_{Tb}^{cal}, for $Tb_3Fe_{29-x}Cr_x$ are also estimated using the following formula on the basis of linearly antiferromagnetic coupling between the magnetic moments of the Tb and Fe sublattices and shown in Table 3.

$$\mu_{Tb}^{cal} = \frac{\left[(29-x)\mu_{Fe}^{MS} - \mu_s^{exp}\right]}{3}$$

We can observe that the average Fe magnetic moments obtained for the $Tb_3Fe_{29-x}Cr_x$ compounds at 300 K also decrease linearly with Cr concentration.

Table 3. The saturation magnetization (M_S), the Mössbauer fitted values of the average hyperfine fields (B_{hf}), and the corresponding average Fe magnetic moments (μ_{Fe}^{MS}) for $Y_3Fe_{27.2}Cr_{1.8}$ and $Tb_3Fe_{29-x}Cr_x$ (x=1.0; 1.5; 2.0; and 3.0) at 300K. The deduced average Tb magnetic moments (μ_{Tb}^{Calc}) for $Tb_3Fe_{29-x}Cr_x$ compounds are listed in the last column.

$R_3Fe_{29-x}Cr_x$	M_S (μ_B/f.u.)	B_{hf} (T)	μ_{Fe}^{MS} (μ_B/Fe)	μ_{Tb}^{Calc} (μ_B/Tb)
$Y_3Fe_{27.2}Cr_{1.8}$	38.6	19.1	1.42	-----
$Tb_3Fe_{28.0}Cr_{1.0}$	27.0	24.3	1.81	7.89
$Tb_3Fe_{27.5}Cr_{1.5}$	22.8	22.6	1.68	7.80
$Tb_3Fe_{27.0}Cr_{2.0}$	22.0	19.2	1.43	5.54
$Tb_3Fe_{26.0}Cr_{3.0}$	17.0	17.2	1.28	5.43

Mössbauer spectra of the $Tb_3Fe_{27.0}Cr_{2.0}$ at 77K, 150K and 200K are shown in Figure 3. Both the saturation magnetization and the ratio of B_{hf} to μ_{Fe} for $Tb_3Fe_{27.0}Cr_{2.0}$ for these temperatures are deduced from their values at 4.2K and 300K, assuming a linear decrease from 4.2K to 300K. We can note (Table 4) that the average Fe and Tb moments decrease with increasing temperature, except in the temperature range from 150K to 200K. Two turning points between 150K and 200K in the termomagnetic curve of $Tb_3Fe_{27.0}Cr_{2.0}$ were observed, which may correspond to the spin reorientation transitions (SRT), reflecting a temperature induced spin phase transition from one spin phase to another. The unusually small values of the average Fe and Tb magnetic moments at 200K were deduced from the formula above. They are due to the non-linearly antiferromagnetic coupling between the magnetic moment of the Tb and Fe sublattices in the temperature range from 150K to 200K.

In conclusion, the average Fe magnetic moments in these compounds obtained from Mössbauer spectroscopy, at 4.2K and 300K, agree with magnetization measurements. The hyperfine fields $B_{hf}(2)$ of Fe_5 and Fe_{10} sites, with the largest R-atom NN numbers (3), are unusually larger, which could indicate a significant spin polarization induced by the neighboring rare earth atoms. The average hyperfine fields of $Tb_3Fe_{29-x}Cr_x$ at 4.2K and 300K decrease linearly with Cr concentration, which is

also in accordance with the variation of the average Fe magnetic moment in the
$Tb_3Fe_{29-x}Cr_x$ compounds.

Table 4. The saturation magnetization (M_s), the Mössbauer fitted values of the average hyperfine fields
(B_{hf}), the ratio χ of the average hyperfine field relative to the average Fe magnetic moment, the
corresponding average Fe magnetic moments (μ_{Fe}^{MS}), and the deduced average Tb magnetic moments
(μ_{Tb}^{Calc}) for $Tb_3Fe_{27.0}Cr_{2.0}$ measured at 4.2K, 77, 150, 200, and 300K

T (K)	M_s (μ_B/f.u.)	B_{hf} (T)	χ (T/μ_B)	μ_{Fe}^{MS} (μ_B/Fe)	μ_{Tb}^{Calc} (μ_B/Tb)
4.2	21.30	26.7	14.50	1.84	9.46
77	21.45	20.9	14.25	1.47	6.08
150	21.65	20.1	14.00	1.44	5.74
200	21.75	19.5	13.80	1.41	5.44
300	22.00	19.2	13.45	1.43	5.54

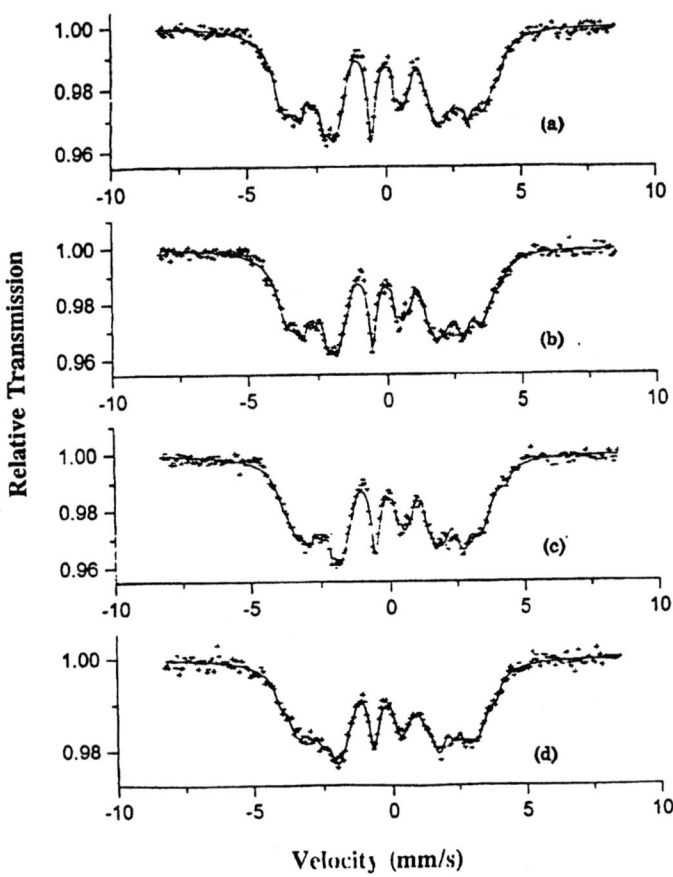

Fig 3.Mössbauer spectra of $Tb_3Fe_{27.0}Cr_{2.0}$ at 77K (a), 150K (b), 200K (c) and 300K (d).

Acknowledgments

This work has benefited from discussions with Dr. Alberto P. Guimarães, and it was supported by the exchange program between CNPq- Brazil and State Committee of Science and Technology – P. R. China.

References

[1] F.M. Yang, B. Nasunjilegal, J.L. Wang, J.J. Zhu, W.D. Qin, N. Tang, R.W. Zhao, B.P. Hu, Y.Z. Wang, and H.S. Li, J. Phys.: Conden. Matter. 7 ((1995) 1679.

[2] B.P. Hu, G.C. Liu, Y.Z. Wang, B. Nasunjilegal, N. Tang, F.M. Yang, H.S. Li, and J.M. Cadogan, J.Phys.: Conden. Matter. 6 (1994) L595.

[3] Y.Z. Wang, B.P. Hu, G.C. Liu, H.S. Li, X.F. Han, and C.P. Yang, J. Phys.: Conden. Matter. 9 (1997) 2793.

[4] D.H. Ryan, J.M. Cadogan, A. Margarian, and J.B. Dunlop, J. Appl. Phys. 76 (1994) 6150.

[5] X.F. Han, R.G. Xu, X.H. Wang, H.G. Pan, T. Miyazaki, E. Baggio-Saitovitch, F. M. Yang, and C.P. Cheng, J. Phys.: Conden. Matter. 10 (1998) 7037.

[6] X.F. Han, H.G. Pan, H.L. Liu, F.M. Yang, and Y.W. Zheng, Phys. Rev. B 56 (1997) 8867.

[7] X.F. Han, F.M. Yang, H.G. Pan, Y.G. Wang, J.L. Wang, H.L. Liu, N. Tang, R.W. Zhao, and H.S. Li, J. Appl. Phys. 81 (1997) 7450.

[8] W.B. Yelow and Z. Hu, J. Appl. Phys. 79 (1996) 1330.

[9] P.C.M. Gubbens, J.H.F. van Apeldoorn, A.M. Vander Kraan, and K.H.J. Buschow, J. Phys. F 4 (1974) 921.

ANTIFERROMAGNETIC PROPERTIES OF Fe$_x$Mn$_{0.95-x}$Al$_{0.05}$, 0.40 ≤ x ≤ 0.70 ALLOYS, IN THE FCC DISORDERED PHASE

G. Medina, G. A. Pérez Alcázar, J. Tabares and A. García.
Departamento de Física, Universidad del Valle, A. A. 25360 Cali, Colombia

In this work we report the experimental studies of the Fe$_x$Mn$_{0.95-x}$Al$_{0.05}$ alloy series in the fcc disordered phase by ac susceptibility, Mössbauer spectroscopy, ac calorimetry at low temperatures, X-ray diffraction and Differential Scanning Calorimetry (DSC). Results show that these alloys are antiferromagnetic in the range from 40 at. % Fe to 70 at. % Fe, but this antiferromagnetic behavior becomes diluted as the Fe concentration is increased. The antiferromagnetic order can be obtained in three possible types of spin structure, each one in the <001>, <110> and <111> directions, respectively. These antiferromagnetic structures depending both, on the temperature and on the composition. Also, it was possible to obtain the Néel temperatures for some compositions.

1. Introduction

Earlier investigations on the γ-FeMn alloys system were carried out elsewhere in order to explore the antiferromagnetism of γ-Fe, taking into account that Fe-Mn alloys are typical iron alloys which are stable at low temperatures. The magnetic properties of Fe-Mn alloys were, however, found to be different from those of γ iron as well as γ manganese. Studies by neutron diffraction at room temperature of the γ-FeMn alloys [1], suggested that they have an orthorhombic spin structure which consists of four different <111> directions. These results are complemented by theoretical studies [2] that propose three types of spin structure that minimize the exchange energy and anisotropy energy of the FCC structure, each one in the <001>, <110> and <111> directions. Chakrabarti [3] has suggested a structural phase diagram, obtained by X-ray diffraction, of the ternary Fe-Mn-Al system for alloys quenched from 1000 °C. The phase diagram shows that the austenitic phase (FCC) is stable for low Al concentrations and below 60 at.% Mn. In order to contribute to the understanding of the magnetic properties of the ternary system γ-FeMnAl with 5 at. % Al, this paper describes the experimental results of the Fe-Mn-Al system in the FCC disordered phase. Our results complement those reported in the literature for γ-FeMn and γ-FeMnAl alloys [4, 5, 6, 7].

2. Experimental

The alloys were prepared in the same way reported by Chakrabarti [3]. After this powdered samples were prepared for Mössbauer measurements, X-ray diffraction, DSC calorimetry, ac susceptibility and thin sheets of 90μ for ac calorimetry. Magnetic measurements were performed with using a Lake-Shore model 7130 susceptometer with H_{ac}=3 Oe, H_{dc}=10 Oe and f =175 Hz, in a temperature range from 12 to 300 K. The X-ray diffraction measurements were made by using a Rigaku 2000 diffractometer. The Mössbauer measurements at low temperature were made by using a conventional multichannel spectrometer model MS1200 with a Co^{57} source and a Janish Research cryostat. Spectra were fitted using the Normos program with 48 sextets.

3. Experimental results

From X-ray diffraction measurements it was possible to show that the alloys are all in the FCC phase with a lattice parameter remains nearly constant to a value of 3.63Å as the Fe concentration increases, as expected for a characteristic FCC structural phase, according to the reports in the literature [3,7].

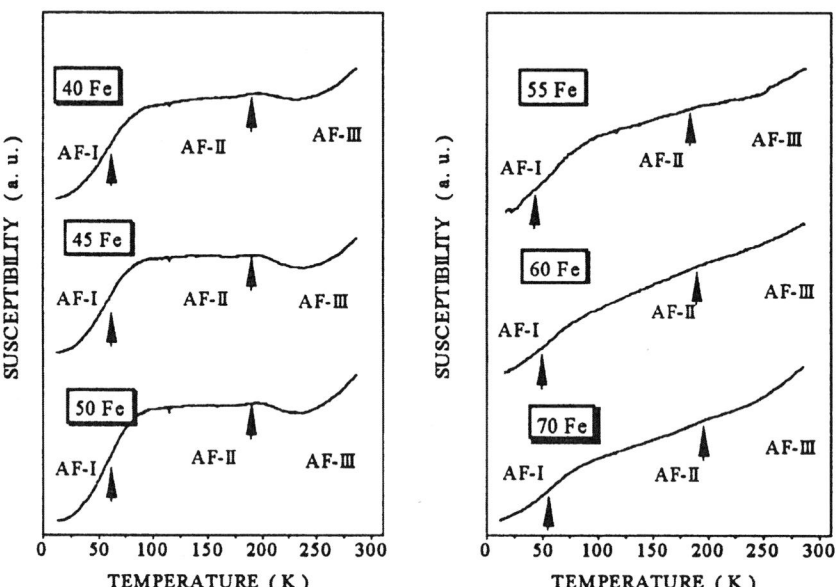

Fig. 1. Magnetic susceptibility versus temperature of FeMnAl alloys with thw f.cc structure.

Fig. 1 shows the ac susceptibility versus temperature curves for the alloy series in a temperature range from 12 to 300K. These curves show a similar ascendant behavior, characteristic of an antiferromagnetic phase, with two observed inflection points approximately at 60 and 180K. These inflection points divide the temperature range in three intervals, which we labeled as AF-I, AF-II and AF-III, corresponding to temperatures before 60K, between 60 and 180K and between 180 and the Néel's temperature, expected above room temperature.

Fig. 2 shows the DSC calorimetry obtained results for Fe concentrations between 55 and 70 at.%. These curves present a change in their baseline, characteristic of a second order transition which is associated to the phase transition from antiferromagnetic to paramagnetic phase, corresponding to the Néel temperatures. These results complement the ac susceptibility results. The ascendant behavior, observed at temperatures higher than 525K, is linked to oxidation effects and by this reason it was not possible to find the Néel temperature for alloys with 40, 45 and 50 at. % Fe respectively, since, according to the present obtained results, they are expected to be in the observed oxidation conditions.

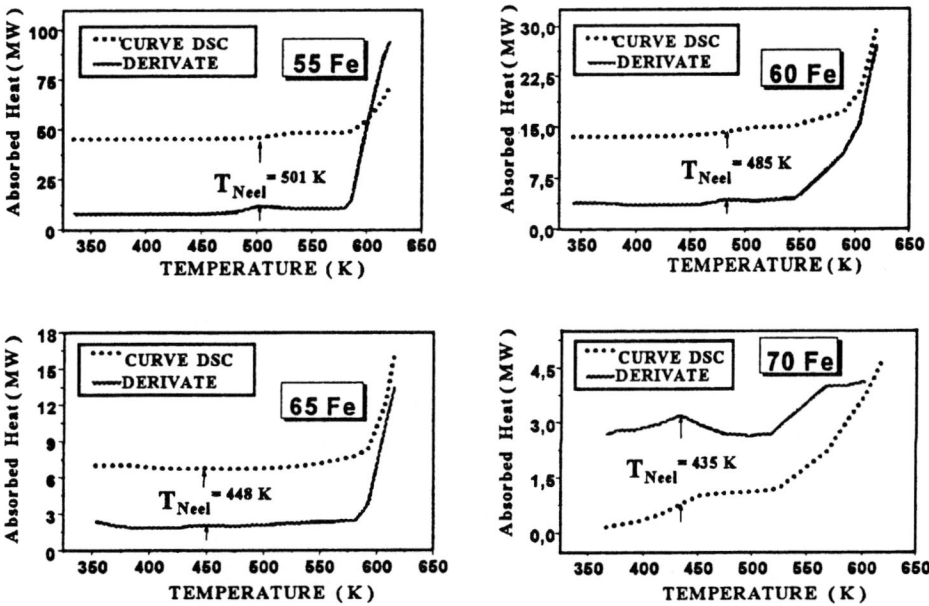

Fig. 2. DSC calorimetry of $Fe_x Mn_{0.95-x} Al_{0.05}$, $0.55 \leq x \leq 0.70$ alloys, at high temperatures. Arrow indicates Néel temperature.

In order to complement the obtained ac susceptibility results, alloys with 45 and 50 at. % Fe were selected in order to accomplish measurements by ac calorimetry and the alloy with 45 at. % Fe to accomplish measurements by Mössbauer spectroscopy, at temperatures between 25 and 275K.

By ac calorimetry, the curves showed in fig. 3 were obtained, in which two anomalies are presented at temperatures near to 80 and 210K, in a similar way to that of ac susceptibility, but for slightly higher temperatures.

Fig. 4 shows the Mössbauer spectra and their corresponding hyperfine field distributions (HFD) obtained for the alloy with 45 at. % Fe at low temperatures,

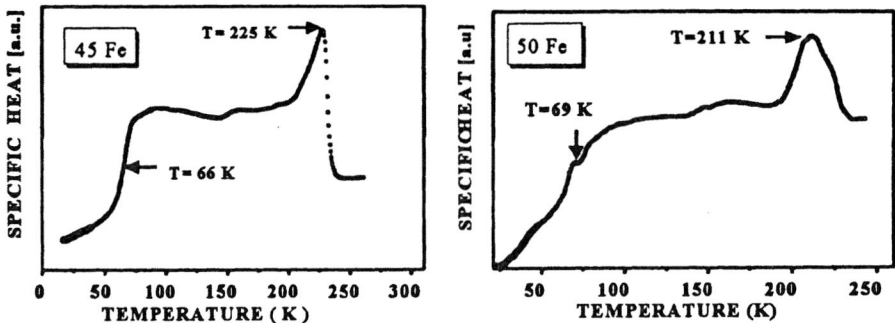

Fig. 3. AC calorimetry of FeMnAl alloys with 45 and 50 at. %Fe, at low
temperatures. Arrow indicates the presence of two anomalies.

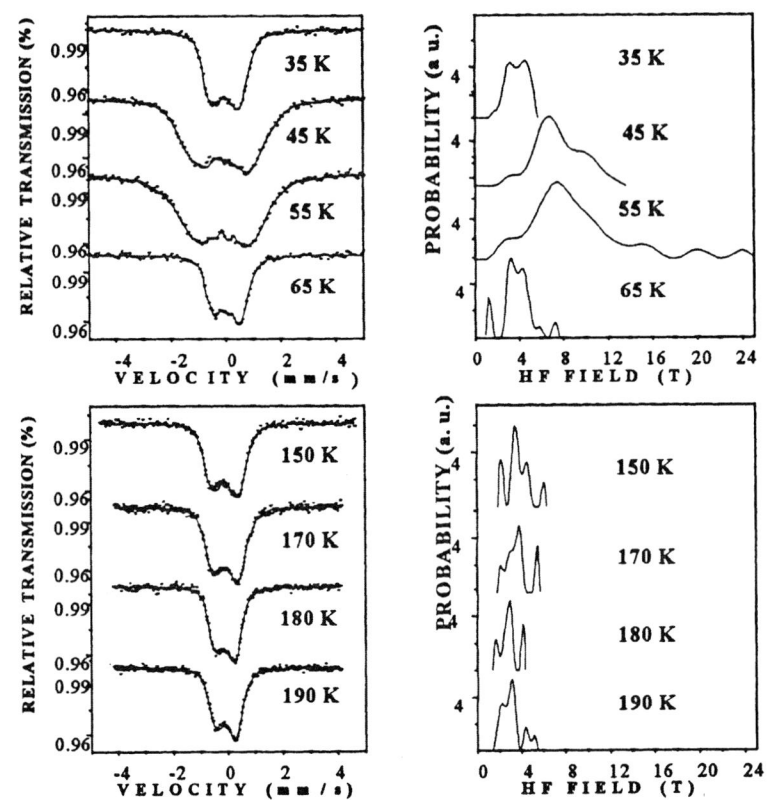

Fig. 4. Mössbauer spectra and their hyperfine field distribution for
Fe$_{0.45}$ Mn$_{0.50}$ Al$_{0.05}$ alloys, at several temperatures.

in the ranges between 35 to 65K and 150 to 210K. These ranges include the
corresponding inflection temperatures found in the susceptibility and calorimetry
measurements. These spectra present a pronounced broadening accompanied

with an increase in the HFD for 45 and 55K, followed by a decrease in the amplitude of the spectrum up to 190K.

Fig. 6 shows the curve of the mean hyperfine field (MHF) versus temperature, in which it is possible to observe a clear increase in the MHF near 55K and a small decrease near 180K. Furthermore, a graph of the line width versus temperature is shown in fig. 5, revealing a pronounced increase between 35 and 65K, and other less pronounced between 100 and 200K. These ranges are in agreement with the critical temperatures of 60 and 180K obtained by ac susceptibility.

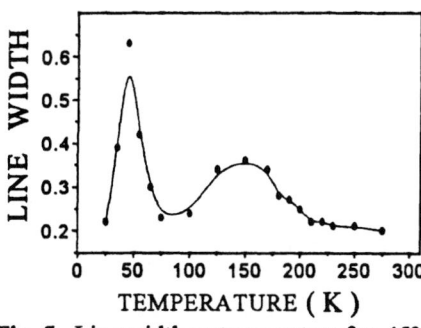
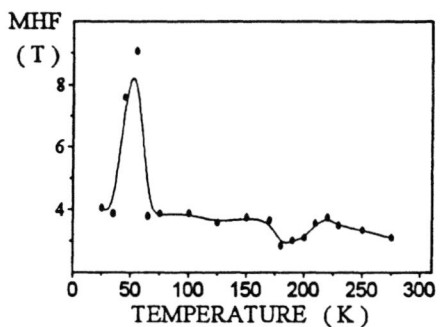

Fig. 5. Line width vs temperature for 45% at. Fe.   Fig. 6. Mean hyperfine field vs temperature for 45% at. Fe.

4. Discussion

The two inflection points obtained by ac susceptibility and verified by ac calorimetry divide the temperature range in these three zones, which we labeled as AF-I, AF-I and AF-III. Furthermore, we associate these three zones to the three types of antiferromagnetic bonds, in the <001>, <110> and <111> directions respectively, in agreement with those theoretically proposed by Umebachi and Ishikawa [2]. These different couplings were obtained by minimizing the exchange and the anisotropy energies for the fcc structure (fig. 7). Coupling in the <001> direction is associated with range AF-I, due to the poor influence of the thermal energy in the interaction of the spins allowing a more anisotropic coupling. Coupling in the <110> direction is associated with range AF–II and that in the <111> direction with AF-III. This is so due to the fact that if the temperature increases, the antiferromagnetic coupling between the spins tends to be aligned in directions less anysotropic, and furthermore, with the increase in T this coupling dissapers and is the responsible for the transition to the paramagnetic phase, such as was reported for the alloys γ-FeMn [4]. In order to complement the ac susceptibility results, the samples with Fe concentrations of 45 and 50 at. %Fe were measured by ac calorimetry and the results are showed in fig. 3, evidencing the same anomalies, but with a shift to bigger temperatures in the limits of the intervals AF - I and AF - II.

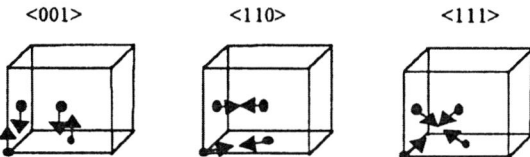

Fig. 7. Stable spin structures which minimize exchange energy and anisotropy
Of the FCC structure, in the <001>, <110>, and <111>.

According to the results obtained by Mössbauer spectroscopy for the alloy with 45 at. % Fe at temperatures in a range from 25 to 275K (fig. 5), it was obtained two strange behaviors not reported in the literature. In the first one it is observed a broadening of the spectrum in the range between 35 to 65K, that coincides with the proposed change in the antiferromagnetic coupling between the <001> direction to the <110> direction, upon increasing thermal energy of the system. This anomaly in the broad of the hyperfine field distribution is better evidenced in the curve of MHF Vs T of the Fig. 6, which presents a clear increase near 55K. In the second anomalous behavior that was found within range of 170 to 190K, it is observed a decrease in the shape of the spectrum, range that coincides with the proposed change of the coupling from the <110> direction to the <111> direction, upon increasing thermal energy of the system. The small decrease of the MHF observed in the fig. 6 about 180 K corroborates the detected small anomaly. Also the increase in the line width with the temperature detected about 60 and 180K (Fig. 5) gives new evidence of the anomalies associated with the changes of coupling within the antiferromagnetic phase proposed. The obtained results will be confronted with neutrons diffraction studies which are now in progress will give the definitive evidence of our proposed model.

5. Acknowledgements

The authors acknowledge support for their work by COLCIENCIAS and Universidad del Valle, Colombia.

5. References

[1] J. S. Kouvel and J. S. Kasper, J. Phys. Chem. Solids 24 (1963) 1743.
[2] H. Umebayashi and Ishikawa, J. Phys. Soc. Japan 21 (1966) 1281.
[3] D. J. Chakrabarti, Met. Trans. B. 8B (1977) 121.
[4] C. Paduani, Doctoral Thesis, Departamento de Física Universidad Federal de Minas Gerais, Bello Horizonte, Brasil, 1986.
[5] Y. Ishikawa and Y. Endoh, J. Phys. Soc. Japan 23 (1967) 205.
[6] Y. Endoh and Y. Ishikawa, J. Phys. Soc. Japan 30 (1971) 1614.
[7] G. A. Pérez Alcazar, and E. Galvao Da Silva, Hyperfine Interactions 66 (1991) 221-230.

MAGNETIC PHASE DIAGRAM OF Ce(Co$_{1-x}$Fe$_x$)Ge$_3$ AS SEEN BY A LOCAL PROBE

D. R. Sánchez[a], S. N. de Medeiros[b], S. L. Bud'ko[c], M. B. Fontes[a], M.A.Continentino[b] and E. M. Baggio-Saitovitch[a].

[a]*Centro Brasileiro de Pesquisas Fisicas (CBPF). Rua Xavier Sigaud 150, Urca, 22290-180, Rio de Janeiro, R.J., Brazil.*
[b]*Universidade Federal Fluminense (UFF). Campus da Praia Vermelha S/N, 24210-340, Niterói, R.J., Brazil.*
[c]*Ames Laboratory and Department of Physics and Astronomy, Iowa State University, Ames, Iowa, 50011, USA.*

The CeFeGe$_3$ crystallizes in the body centered BaNiSn$_3$ structure and is a moderate heavy fermion compound which does not exhibit any magnetic phase transition above 0.5 K. The CeCoGe$_3$ is also a heavy fermion, exhibiting a high Sommerfeld's coefficient value (γ = 150mJ/molK^2) and presenting a complex magnetic phase diagram. In this work we use ^{57}Fe Mössbauer spectroscopy (MS) and ac magnetic susceptibility measurements as a function of the temperature in order to elucidate the phase diagram of the Ce(Co$_{1-x}$Fe$_x$)Ge$_3$ series of compounds and search for a quantum critical point in the Co rich side of the series.

1- Introduction:

The family of *CeTMGe$_3$* (TM = transition metal) compounds crystallizes in the body centered *BaNiSn$_3$* structure (Fig.1). This structure is related to the *ThCr$_2$Si$_2$* structure, where a great numbers of materials display exotic physical properties, i.e., intermediate valence (IV), heavy fermions (HF), magnetism and superconductivity, complex magnetic phase diagrams.

The *CeFeGe$_3$* [1] is known as moderate heavy fermion (HF) material ($\gamma \approx$ 150 mJ/mol K^2) with high Kondo temperature, no magnetic order down to 0.5 K and well defined integral (*Ce^{3+}*) valence of *Ce* at high temperatures. The *CeCoGe$_3$* was reported [2] as HF ($\gamma \approx$ 111 mJ/molK^2) with complex low temperatures magnetic phase diagram with two magnetic transitions at ~21 K and ~18 K in the absence of a magnetic field. The upper ordering temperature is believed to be a transition from the paramagnetic state to a c-axis ferrimagnetic state, which in turn transforms into a collinear antiferromagnetic state [2] at ~ 18 K.

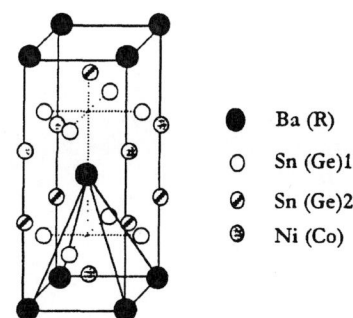

Fig.1: The *BaNiSn$_3$* crystalline structure.

The magnetic properties of HF systems are governed by a delicate balance between *RKKY* interactions and the *Kondo* effect. The competition between these two mechanisms can be treated using the *Doniach* model [3]. The important parameter in this model is *J/W*, where *W* is the width of the conduction band and *J* is the exchange integral between conduction electrons and localized *f* moments. The parameter *J/W* can be varied by chemical substitution or hydrostatic pressure. This model suggests that the change from an antiferromagnetic ordering to a *Kondo*-spin compensated ground state at zero temperature is a second order transition, and occurs gradually as the parameter *J/W* reaches a critical value (*J/W*)$_c$, named Quantum Critical Point (QCP). Applying the scaling theory of critical phenomena, *Continentino et al.* [4] have shown the existence of a coherence line on the non critical side of the phase diagram of the *Kondo* lattice ((*J/W*) > (*J/W*)$_c$), associated with the QCP. The coherence temperature (*T$_{coh}$*) provides the characteristic temperature, at which the system enters into the renormalized *Fermi* liquid regime. For $(J/W) < (J/W)_c$ the characteristic temperature is the *Néel* (or *Curie*) temperature, which marks the appearance of long range magnetic order. At the critical point, i.e., $(J/W) = (J/W)_c$, we expect to find *non-Fermi-liquid* behavior down the lowest temperatures since the system does not cross the coherence line.

The search for the QCP and the modification of the magnetically ordered state in the Co-rich compounds with Fe-concentration in the $Ce(Co_{1-x}Fe_x)Ge_3$ series are the main motivation of our work. The decrease of the unit cell volume from $CeFeGe_3$ to $CeCoGe_3$ is only 1.5 %. Since the change in the unit cell volume between the end compounds is very small, and having in mind that the suggested substitution is non - isoelectronic, the change of the number of electrons in the conduction band is expected to be the main mechanism responsible for the changes in the ground state. The study of this series has been performed using several techniques, like Mössbauer spectroscopy, DC magnetization, AC susceptibility and magnetoresistivity. However, in this communication we have chosen to present mainly our results on Mössbauer spectroscopy. We have shown that ^{57}Fe Mössbauer effect (ME) spectroscopy can be a powerful tool to study the magnetic structure (via transferred magnetic field at the *Fe* site) in systems with the $ThCr_2Si_2$ structure [5,6] wherein the *Fe* does not carry magnetic moment.

2- Experimental

Polycrystalline samples of $CeFeGe_3$ and $CeCoGe_3$ were prepared by conventional arc-melting of the stoichiometric amounts of high purity elements with subsequent annealing at 750°C for 5 days. The X - ray diffraction patterns reveal that single phase compounds with no impurity peaks were formed. The ^{57}Fe Mössbauer measurements were performed in transmission geometry and constant acceleration mode, with a 25 mCi ^{57}Co in α–Rh matrix. The spectra were measured as a function of temperature with the source kept at ambient temperature. The central shift (CS) values are given relative to α-Fe.

3- Results and Discussion

In the Fig.2 we present the AC susceptibility measurements (χ_{AC}) for the $Ce(Co_{1-x}Fe_x)Ge_3$; x-Fe = 0, 0.02 and 0.6, samples. The arrow indicates the antiferromagnetic transition temperature (T_N) for x-Fe = 0. The magnetic transition at 21 K can not be observed in the χ_{AC} measurements, as noticed by Pecharsky et al. [2]. For x-Fe = 0.02 we observe a sharp peak around 14 K. The understanding of the nature of this magnetic transition is among the aims of our work. The AC susceptibility curve for x-Fe = 0.6 does not show any magnetic transition. The negative points in the curves (see Fig.2) are due to the diamagnetic contribution of the sample holder.

We have performed ^{57}Fe Mössbauer measurements of $Ce(Co_{1-x}Fe_x)Ge_3$ (x-Fe = 0.02 and 0.6). Above 20 K the spectra show a single quadrupole doublet (Figs.3 and 4) for both samples, with respective values ΔE_Q = 0.39mm/s and ΔE_Q = 0.32mm/s, which is attributed to *Fe* in the $BaNiGe_3$ structure.

In the temperature dependent Mössbauer spectra for x-Fe = 0.02 one can observe a gradual line broadening and the appearance of an asymmetry with the decrease of temperature, which can be interpreted as a signature of a small hyperfine field B_{hf}. Below 17 K the spectra were fitted using the complete hamiltonian, assuming a magnetic and quadrupole interactions of the same order of magnitude. The best fits were obtained for the angle between the z-component of the electric field gradient and the hyperfine field equal zero, implying that the moments are along the c-axis. We show in the Fig.5 the dependence of the B_{hf} with the temperature for x-Fe = 0.02. Since *Fe* atoms do not carry magnetic moment, this B_{hf} is believed to be transferred via RKKY interactions from the ordered neighboring *Ce* moments, to the ^{57}Fe site.

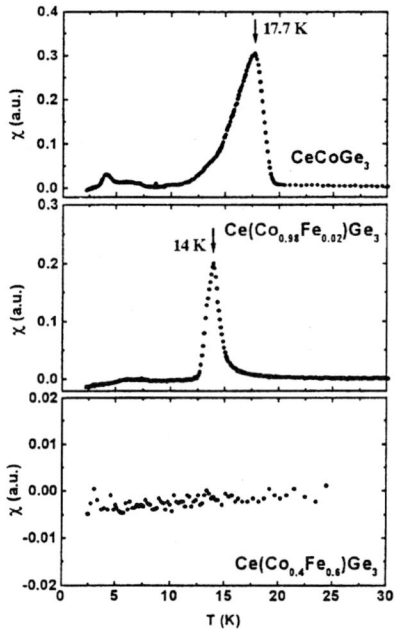

Fig.2: AC susceptibility curves for the $Ce(Co_{1-x}Fe_x)Ge_3$; x-Fe = 0, 0.02 and 0.6, samples.

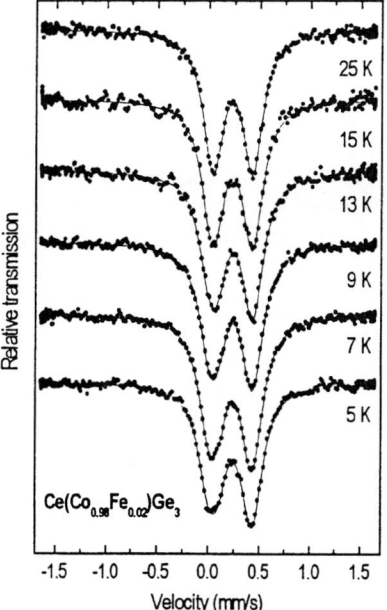

Fig.3: Temperature dependence of the Mössbauer spectra for the $Ce(Co_{0.98}Fe_{0.02})Ge_3$ sample.

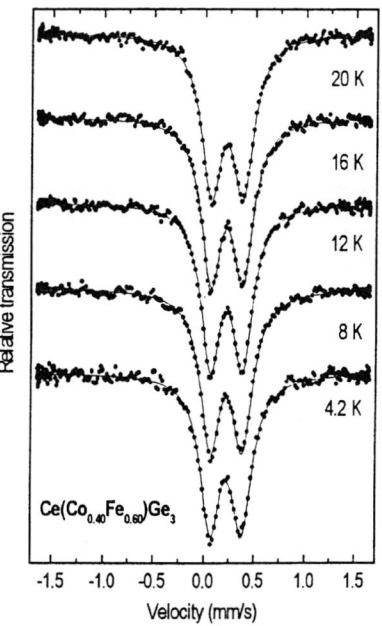

Fig.4: Temperature dependence of the Mössbauer spectra for the $Ce(Co_{0.4}Fe_{0.6})Ge_3$ sample.

Supposing that the magnetic transition observed for the x-Fe = 0.02 sample is still antiferromagnetic, one could understand the presence of a transferred field at the *Fe* site, inspecting at its position in the crystalline structure. The *Fe* atom sits inside a pyramid formed by *Ce* atoms, near its base. Therefore, the transferred hyperfine field does not cancel completely at the *Fe* site, even for a collinear antiferromagnetic order, leading to this small measured field. Any other kind of magnetic arrangement would also result in an observable transferred field, possibly with higher intensity.

For the x-Fe = 0.6 sample, the Mössbauer spectra do not show any change with temperature, in fair agreement with the AC susceptibility measurements (see Fig.4).

Mössbauer measurements as a function of the temperature for the complete series is still in progress, in order to follow the change of the magnetism with *Fe*-concentration.

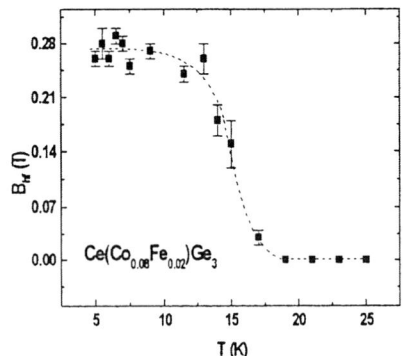

Fig.5: Temperature dependence of the B_{hf} obtained for the $Ce(Co_{0.98}Fe_{0.02})Ge_3$ sample.

4- Conclusion

We have shown our preliminary results on Mössbauer spectroscopy and AC susceptibility of the $Ce(Co_{1-x}Fe_x)Ge_3$ (x-Fe = 0, 0.02 and 0.6) series of compounds. The temperature dependence of the Mössbauer spectra for x-Fe = 0.02 and 0.6 is in good agreement with the AC susceptibility measurements, concerning the magnetic order. The nature of the magnetism for x-Fe = 0.02 is still not completely clear, since the position of *Fe* atoms in the lattice favors a non-zero transferred field for any kind of spin arrangement. From Mössbauer measurements the magnetic moments are probably aligned along the c-direction in an antiferromagnetic configuration. However, more studies are in progress in order to understand the modification of the magnetic properties of $CeCoGe_3$ system introduced by *Fe*-doping

Acknowledgements: This work was supported by FAPERJ and CNPq (Brazil).

References:

[1] H.Yamamoto et al., Phys. Lett. A **196** (1994) 83.
[2] V.K.Pecharsky, O.–B.Hyun, and K.A.Gschneidner, Jr., Phys. Rev B **47**, 11839 (1993).
[3] S.Doniach in: "*Valence Instabilities and Related Narrow Band Phenomena*", ed. R.D.Parks, (Plenum, NY, 1977) p.169; S.Doniach, Physica B **91** (1977) 231-234
[4] M.A.Continentino, G.M.Japiassu and A.Troper, Phys. Rev. B **39** (1989) 9734 / M. A. Continentino, J. Phys. I, **1** (1991) 693.
[5] E.Baggio-Saitovitch, in this Conference.
[6] D.Sanchez, H.Micklitz, M.B.Fontes, S.L.Bud'ko and E.M.Baggio-Saitovitch, Phys. Rev. Lett., **76** (1996) 507-510.

Mössbauer study of the silicate phase occurring in a calcite marble of Andranondambo, South Madagascar.

J. De Grave[1], S.G. Eeckhout[2‡], E. De Grave[2†], R. Vochten[3] and Ph. de Parseval[4]

[1]Department of Geology and Soil Science, University of Gent, B-9000 Gent, Belgium
[2]Department of Subatomic and Radiation Physics, University of Gent, B-9000 Gent, Belgium
[3]Department of Chemistry, University of Antwerp, B-2020 Antwerpen, Belgium.
[4]Electronic Microprobe Service, Paul-Sabatier University, F-31062 Toulouse, France

An Al-Ca-rich silicate phase in a calcite matrix was sampled in the high-grade metamorphic terrain of the Precambrian Androyen complex of SE Madagascar. It consists of black and brownish crystals, between which no drastic structural or compositional differences were detected. X-ray diffraction indicates a diopside structure, and from wavelength-dispersive microprobe analyses an average composition as $(Ca_{0.98}Na_{0.03})\{Mg_{0.68}Al_{0.16}Fe_{0.10}Ti_{0.04}\}[Si_{1.77}Al_{0.23}]O_6$ was found. Mössbauer spectra for a black crystal were collected at various temperatures between 11 K and 470 K. The spectra are adequately described by a superposition of a ferrous and a ferric quadrupole-splitting distribution. The characteristics of the calculated probability distributions are discussed in terms of compositional fluctuations. For comparison, room-temperature spectra for two ferrian aluminian diopsides ("fassaites") are also presented.

A metasomatic Al-Ca-rich silicate phase in a calcite matrix was sampled in the high-grade metamorphic terrain of the Precambrian Androyen complex of SE Madagascar, near the village of Andranondambo. The silicate phase consists of thermally affected black and brownish crystals, between which no drastic structural or compositional differences were detected, using various mineralogical techniques. For instance, X-ray diffraction patterns of the two types of crystals are identical and were identified using the ZDS-system© search routine as arising from a diopside structure. The reflections are quite sharp, implying a high degree of crystallinity. Wavelength dispersive microprobe analyses, using proper reference materials, on nine spots (spot size 2x2 μm) of a thin section, have revealed only minor fluctuations in the elemental compositions on a micro scale. The average composition, calculated on the basis of six oxygens and with the requirement of complete charge balance, can be written as:

$$(Ca_{0.98}Na_{0.03})\{Mg_{0.68}Al_{0.16}Fe^{2+}_{0.07}Fe^{3+}_{0.03}Ti_{0.04}\}[Si_{1.77}Al_{0.23}]O_6$$

‡ Research Assistant, Fund for Scientific Research - Flanders
† Research Director, Fund for Scientific Research - Flanders

where the various brackets represent the M2, M1, and tetrahedral sites respectively. According to the most recent nomenclature conventions for pyroxenes [1], this phase is a ferrian aluminian diopside, formerly called *fassaite*.

Transmission Mössbauer spectra (MS) for a powdered black crystal, sample MA2, were collected at various temperatures between 11 K and 470 K using standard equipment. Example spectra ($T = 125$ and 470 K) are reproduced in Fig.1. MS at room temperature (RT) recorded for a second black crystal and for a brown crystal were almost identical to that of sample MA2. The spectra clearly are composed of broad Fe^{2+} and Fe^{3+} doublets and rather resemble those obtained for certain Fe-bearing silicate glasses [2] than those for diopside [3] or ferrian diopside [4]. However, since no glassy state was observed by optical microscopy on thin sections, the presence of iron in a glass matrix may be ruled out. The considerable line broadening may therefore be attributed to local compositional fluctuations in the cationic configurations in the nearest-neighbour co-ordinations of the respective iron probes. The complex chemical formula indicated above, with a considerable Al-for-Si substitution, is consistent with this qualitative interpretation.

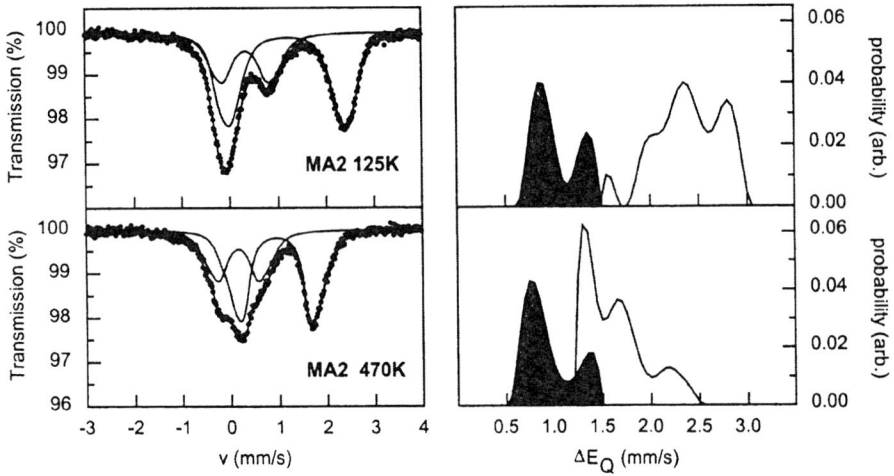

Fig.1. Mössbauer spectra of sample MA2 at 125 K and at 470 K (left), and calculated probability distributions for the ferrous and ferric (shaded area) quadrupole splittings.

A superposition of a ferrous and a ferric model-independent quadrupole-splitting distribution (QSD) was found to produce adequate fits to the experimental line shapes. A correlation between the quadrupole splitting ΔE_Q and the centre shift δ was found to be insignificant. The calculated probability distributions, $p(\Delta E_Q)$, for the earlier mentioned example MS are shown in the right half of Fig.1. With a very few exceptions only, the multi-modal shape for both the Fe^{2+} and Fe^{3+} $p(\Delta E_Q)$ profiles is observed for all temperatures between 125 K and 470 K. At 100 K and lower, the bi-modal structure of the ferric QSD does not appear. Instead, a more or less symmetric profile results from the calculations. The multi-modal character of the ferrous doublet, with commonly three maxima, remains. It should be noted that the same absorber has been measured in three different cryostat systems and that a total of nine MS in the range 11 – 125 K have been considered, all consistently producing a single-peaked $p(\Delta E_Q)$ for Fe^{3+}.

Some numerical characteristics of the distribution profiles are listed in Table 1 for selected temperatures. These are for Fe^{3+} the centre shift δ (referred to α-Fe), the quadrupole-splitting values of the two maxima when resolved, $\Delta E_{Q,m1}$ and $\Delta E_{Q,m2}$, and the average quadrupole splitting $\Delta E_{Q,av}$. Since the various maxima for the Fe^{2+} profiles are mostly ill defined, only the splitting of the most prominent component, $\Delta E_{Q,m}$, is indicated in Table 1.

Table 1. Numerical characteristics (see text) of the ferric and ferrous quadrupole-splitting distributions fitted to the MS of the Fe-Al diopside MA2 at selected temperatures (upper part) and of the two type-"fassaites" RA613 and RA615 at 295 K (lower part). A is the fractional area of the ferrous component.

	Fe³⁺				Fe²⁺			
T (K)	δ (mm/s)	$\Delta E_{Q,m1}$ (mm/s)	$\Delta E_{Q,m2}$ (mm/s)	$\Delta E_{Q,av}$ (mm/s)	δ (mm/s)	$\Delta E_{Q,m}$ (mm/s)	$\Delta E_{Q,av}$ (mm/s)	A
11	0.57	-	-	0.80	1.26	2.71	2.53	0.68
11*	0.47	-	-	1.01	1.28	2.72	2.48	0.68
80	0.53	-	-	0.83	1.27	2.60	2.56	0.61
80*	0.48	0.77	1.37	0.98	1.30	2.51	2.49	0.68
125	0.43	0.87	1.35	1.03	1.29	2.36	2.40	0.66
225	0.41	0.90	1.40	0.96	1.21	2.19	2.18	0.64
295	0.35	0.89	1.38	0.96	1.16	1.97	2.00	0.63
400	0.31	0.88	1.38	0.95	1.09	1.64	1.77	0.61
470	0.26	0.77	1.37	0.94	1.05	1.32	1.62	0.62
295 (RA613)	0.37	0.90	1.35	0.97	1.17	1.88	1.93	0.30
295 (RA615)	0.38	0.68	1.34	0.87	1.14	1.87	2.14	0.18

*δ for Fe^{3+} fixed in the iteration

It is tempting to ascribe the different maxima in the $p(\Delta E_Q)$ profiles to distinct sites for the involved cations. In ferrian diopside Fe^{3+} is found on M1 and Si sites and has ΔE_Q values of respectively 0.97 and 1.65 mm/s (RT, sample FD10 in Ref. [4]) and centre shifts 0.37 and 0.20 mm/s. These data, in particular the latter one, seem to rule out a four-fold co-ordination for the ferric species in the present silicate phases. It is then more plausible to explain the maxima in $p(\Delta E_Q)$ by the occurrence of a distinct number of essentially different cation configurations on the six Si tetrahedra surrounding the M1 sites that contain the iron probes. Replacing in one of these the Si by Al (probability ~36%) creates a strongly asymmetric charge distribution around the M1 position and hence a higher ΔE_Q for Fe^{3+} at that position. However, the disappearance of the bimodal structure of the ferric $p(\Delta E_Q)$ at temperatures below ~125 K is hard to reconcile with this qualitative explanation. Moreover, several hyperfine parameters, in particular those for Fe^{3+}, seem to undergo sudden changes in the region 100 – 125 K which are hard to explain. It was at first believed that these features were an artefact of the fitting. Fixing the Fe^{3+} centre shifts at values previously found for a natural aegirine sample with composition $Na_{1.06}Ca_{0.06}Mg_{0.04}Fe_{1.01}Al_{0.06}Si_{1.91}O_6$ [6], restores the bimodal shape of the ferric distribution profile at all temperatures T < 125 K, except 11 K, and produces more regular variations of the other Mössbauer parameters. However, the goodness-of-fit χ^2 significantly increases as compared to the unconstraint fits. This finding implies that the results of this fitting procedure might be doubted. The low-temperature behaviour of the MS of the involved ferrian aluminian diopside therefore remains puzzling and may be the subject of further investigations.

The shape of the ferrous $p(\Delta E_Q)$ can also qualitatively be explained by different nearest-neighbour cation arrangements. However, the origin of the three apparent maxima is unclear. The maximum-probability splitting $\Delta E_{Q,m}$ for Fe^{2+} corresponds well with the ferrous M1 splitting in ferrian diopside (1.92 mm/s for FD10 at RT), which might, however, be fortuitous. It should further be noted that the large magnitudes of the ferric ΔE_Q values and the rather steep temperature variations of the ferrous ones, are indicative for drastic deformations of the M1 symmetry in the sample's pyroxene structure. Considering the irregular shape of the ferrous $p(\Delta E_Q)$ profiles and hence the doubtful physical meaning of the average and maximum-probability splittings, it was decided that further interpretation of their temperature variations in terms of the static-crystal-field model had no sense.

An important result concerns the relative spectral contributions of the two charge states. Averaged over eight different temperatures in the range 125 – 295 K, one finds a value of 0.65 ± 0.01 for the Fe^{2+} fractional area (quantity A in Table 1). Taking into account that the average ratio of the Mössbauer fraction of ferrous ions against that of ferric ions in several other pyroxenes is close to 0.88 [5], it is straightforward to calculate the Fe^{2+}/Fe_{tot} ratio as 0.68, which is in remarkable agreement with the above indicated chemical composition as obtained from microprobe analyses. This finding illustrates that this latter technique, if applied as meticulously as in the present work, is a powerful tool for chemical analysis.

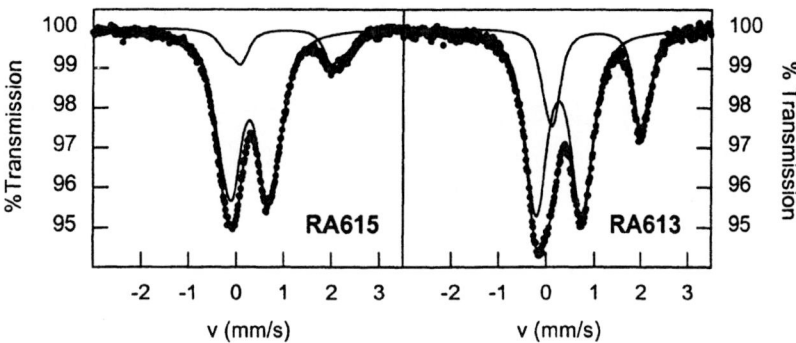

Fig. 2. Room-temperature Mössbauer spectra of two "fassaite" samples from the type locality Val di Fassa, Italy.

To the best of the authors' knowledge, no Mössbauer data on ferrian aluminian diopside or "fassaite" have been reported so far. In an attempt to further identify the marble silicate as a ferrian aluminian diopside, two non-characterised "fassaite" samples, RA613 and RA615, from a museum collection were examined at RT. They have been sampled from the type locality Val di Fassa, Italy. The MS are reproduced in Fig. 2 and, apart from the Fe^{2+}/Fe^{3+} ratio, closely resemble those obtained for the Madasgacar samples. The broad $p(\Delta E_Q)$ profiles also exhibit two or three maxima, however less clearly resolved than in the cases shown in Fig. 1. The relevant Mössbauer parameters for the two mineral species are also included in Table 1 and are significantly different from one another and from those of the MA2 sample. These results are not a sound confirmation of the silicate's nature as a ferrian aluminian diopside, but, considering the possibly broad variations in compositions of the minerals and the high sensitivity of the hyperfine parameters to the composition, the data are not exclusive either.

In conclusion, the results of this Mössbauer study are consistent with the conclusions from mineralogical studies concerning the nature and the composition of the involved silicate phase occurring in the marbles of Andranondambo. They have also demonstrated that the presence of significant amounts of Al in the silicon tetrahedra of the pyroxene structure has a tremendous effect on the shape of the Mössbauer spectra, producing quadrupole-splitting distributions which are many times broader than those resulting from substitutions on the M1 and/or the M2 sites.

Acknowledgements

The authors wish to thank Dr. M. Deliens from the Royal Belgian Institute of Natural Sciences, Brussels, for providing the fassaite samples. This work was supported by the Fund for Scientific Research – Flanders.

References

1. N. Morimoto, Am. Mineral. **73** (1988) 1123
2. E. De Grave and P. Van Iseghem, in: G.J. Long and J.G. Stevens (eds), *Industrial Applications of the Mössbauer Effect*, Plenum Press, New York, 1986, 423
3. W.A. Dollase and W.I. Gustafson, Am. Mineral. **67** (1982) 311
4. E. De Grave and A. Van Alboom, in: I. Ortali (ed.), *Conference Proceedings*, Vol. 50, Societa Italiana di Fisica, Bologna, 1996, 729
5. E. De Grave and A. Van Alboom, Phys. Chem. Minerals **18** (1991) 337
6. E. De Grave, A. Van Alboom and S.G. Eeckhout, Phys. Chem. Minerals **25** (1998) 378

"τ PHASE IN THE QUATERNARY SYSTEM Fe-Ni-B-Sn"

M. A. Ureña, B. Arcondo and H. Sirkin.

Facultad de Ingeniería, Universidad de Buenos Aires. Paseo Colón 850 (1063) Buenos Aires, Argentina.

Previous investigations on the structural evolution from the amorphous to the crystalline state in the $Fe_{38.7}Ni_{38.7}B_{17.4}Sn_{5.2}$ system suggested the presence of a τ phase with a $Cr_{21}W_2C_6$ (D84) type structure in the Fe-Ni-B-Sn quaternary system. This structure is stable in the ternary systems Ni-B-Sn ($Ni_{21}B_6Sn_2$) and Fe-Ni-B ($Ni_{18.5}Fe_{4.5}B_6$), and has not been observed previously in the $(Fe_xNi_{100-x})_{21}B_6Sn_2$ stoichiometry. In order to confirm the existence of a τ phase in the quaternary system, and with the aim to characterise it, alloys of compositions $(Fe_xNi_{100-x})_{21}B_6Sn_2$ were prepared and analysed by X-ray diffraction, Mössbauer spectroscopy and scanning electron microscopy with microprobe facilities.

INTRODUCTION

The τ phases in carbides have the $Cr_{23}C_6$ (D8$_4$) structure type. Unlike the carbides, borides cannot have this structure without a stabilising atomic species such as Sn or Sb at a special lattice position, which is accompanied by an appreciable reduction of atomic size [1]. The structure of the τ (Ni_{21} Sn_2 B_6) phase has four formula units in the face-centred cubic unit cell and the compound contains Sn at the corners of a cube whose edge is half the unit cell edge in length. The Ni_{21} B_6 Sn_2 τ phase was investigated as a possible Mössbauer source because of its particular structural characteristics [2].

Several τ phases are known, such as the stable Ni_{21} Sb_2 B_6 [3] and $Ni_{21}In_2B_6$ [4] phases and the metastable $Fe_{23}B_6$ and $Fe_{4.5}Ni_{18.5}B_6$ [5] phases.

Investigations on the glass forming ability of the Ni-B-Sn ternary system [6,7,8] reveal that the range of compositions in which the system is able to amorphise, extends through the $Ni_{21}Sn_2B_6$ primary field, joining the two ranges of glass forming ability of the Ni–B binary system. This fact may be related to the difficulty of formation of the ternary $Ni_{21}Sn_2B_6$ τ phase whose tin environment differs from that present in the amorphous state.

The structural evolution from the amorphous to the crystalline state in samples annealed at higher temperatures than 470°C for the $Fe_{38.7}Ni_{38.7}B_{17.4}Sn_{5.2}$ system [9] shows a set of X ray diffraction lines that are coincident with the $Ni_{21}Sb_2B_6$ phase, suggesting the presence of a τ phase in the Fe-Ni-B-Sn quaternary system.

In similar investigations in a Metglas 2826 MB alloy (Fe_{40} Ni_{38} Mo_4 B_{18}), *Mizgalski et al* [10] report the existence of the $Fe_xNi_{23-x}B_6$ τ phase crystallising in the metallic glass which disappears when heated (metastable phase). On the contrary, in the $Fe_{38.7}Ni_{38.7}B_{17.4}Sn_{5.2}$ system, the τ phase exists even after heating at a later stage. This fact led to the conclusion of the stability of the phase τ in the Fe-Ni-B-Sn system.

In order to confirm the existence of a τ phase in the Fe-Ni-B-Sn system, and with the aim to characterize it, alloys of compositions $(Fe_xNi_{100-x})_{21}Sn_2B_6$ were prepared and analysed by X-ray diffraction, Mössbauer spectroscopy and scanning electron microscopy with microprobe capabilities.

EXPERIMENTAL PROCEDURE

The samples were prepared using 99.99% purity materials. The alloys were melted in a high frequency furnace under an Argon atmosphere in quartz crucibles previously evacuated to 2.10^{-5}mbar. The cooling was carried out by quenching in air.

The sample compositions were $(Fe_x Ni_{(100-x)})_{21} Sn_2 B_6$ with x = 5, 15, 17, 19.5, 25, 30 (Table I).

The alloys were analysed by optical microscopy, electron probe microanalysis (EPMA) with a wavelengh-dispersive spectrometer (WDX) as well as by X-ray diffraction (Cu Kα radiation) (XRD) and Mössbauer spectroscopy. The Mössbauer experiments (ME) were performed by using a ^{57}Co(Rh) source in transmission geometry at room temperature. Isomer shift (IS) values are referred to Fe(Rh) and IS as well as the quadrupole splitting (QS) values are given with an error of ±0.01.

In order to characterise the τ phase by XRD and ME, a stabilisation annealing at 810°C for 50 hours was carried out in the case of sample T5%.

Because of the difficulty of EPMA quantitative determination of boron, only the Ni/Fe ratio and the identification of the present elements were considered.

RESULTS AND DISCUSSION

Optical microscopy, XRD and EPMA

Optical microscopy revealed in all the samples the presence of a primary phase precipitating with a morphology corresponding to a τ phase [3,6], as well as a small amount of a multiphase constituent in the grain boundaries. The amount of this segregation increased as the Fe content grew, being actually very low in sample T5%. The XRD pattern of this sample (Fig. 1) shows a set of peaks corresponding to the τ phase plus one peak from Ni_3Sn_2. EPMA points out the presence of Fe either in the τ phase or in the Ni_3Sn_2.

Table I: Sample Compositions

SAMPLE	COMPOSITION
T5%	$(Fe_5Ni_{100-5})_{21}Sn_2B_6$ $Fe_{3.64}Ni_{68.66}Sn_{6.9}B_{20.81}$
T15%	$(Fe_{15}Ni_{100-15})_{21}Sn_2B_6$ $Fe_{9.1}Ni_{63.31}Sn_{6.9}B_{20.69}$
T17%	$(Fe_{17}Ni_{100-17})_{21}Sn_2B_6$ $Fe_{12.26}Ni_{60.14}Sn_{6.9}B_{20.69}$
T20%	$(Fe_{20}Ni_{100-20})_{21}Sn_2B_6$ $Fe_{14.17}Ni_{58.23}Sn_{6.9}B_{20.7}$
T25%	$(Fe_{25}Ni_{100-25})_{21}Sn_2B_6$ $Fe_{18.1}Ni_{54.3}Sn_{6.9}B_{20.7}$
T30%	$(Fe_{30}Ni_{100-30})_{21}Sn_2B_6$ $Fe_{21.72}Ni_{50.68}Sn_{6.9}B_{20.7}$

Most of the peaks observed in T5% XRD pattern agree with those reported for $Ni_{21}B_6Sb_2$. In order to identify and characterize the τ phase in the $(Fe_x Ni_{(100-x)})_{21} Sn_2 B_6$ system, its XRD pattern was simulated and the interplanar distance (d_{hkl}) was obtained by using the CaRIne Crystallographic Software.

Table II: Diffraction data and simulation results for sample T5% $(Fe_5Ni_{95})_{21}Sn_2B_6$

(hkl)	2theta exp	2theta sim	d_{hkl} exp (Å)	d_{hkl} sim (Å)
(2 0 0)	16.63	16.71	5.332	5.306
(2 2 0)	23.65	23.72	3.763	3.752
(4 0 0)	33.76	33.79	2.655	2.653
(4 2 0)	37.87	37.92	2.376	2.373
(4 2 2)	41.67	41.70	2.167	2.166
(5 1 1),(3 3 3)	44.36	44.36	2.042	2.042
(4 4 0)	48.50	48.53	1.877	1.876
(5 3 1)	50.88	50.91	1.795	1.794
(4 4 2),(6 0 0)	51.69	51.69	1.768	1.769
(6 2 0)	54.69	54.71	1.678	1.678
(5 3 3)	56.88	56.90	1.619	1.618
(6 2 2)	57.61	57.62	1.600	1.6
(7 1 1),(5 5 1)	62.49	62.51	1.486	1.486
(6 4 2)	65.86	65.87	1.418	1.418
(8 0 0)	71.08	71.07	1.326	1.326
(8 2 0),(6 4 4)	73.63	73.61	1.287	1.287
(6 6 0),(8 2 2)	76.13	76.12	1.250	1.251
(7 5 1),(5 5 5)	77.98	77.98	1.225	1.225
(8 4 0)	80.94	81.05	1.188	1.186
(7 5 3),(9 1 1)	82.92	82.89	1.164	1.165
(6 6 4)	85.98	85.93	1.131	1.131
(9 3 1)	87.77	87.74	1.112	1.112

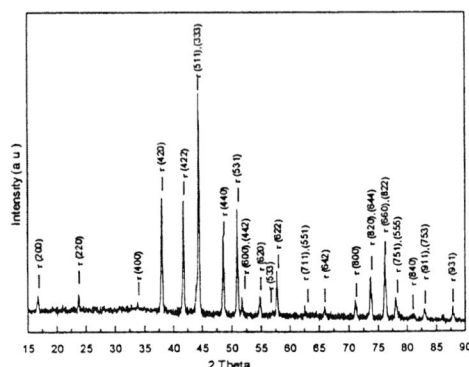

Fig. 1: T5% XRD pattern (Cu Kα radiation)

The space group of the τ phase unit cell is *Fm3m* and the simulation was made using the atomic positions given for the $Ni_{21}Sb_2B_6$ locating Sn in the Sb site and considering an occupational factor of (100-x)/100 for the Ni positions and replacing the other x/100 for the Fe [11].

Table II shows the diffraction data for the cubic boride phase $(Fe_xNi_{100-x})_{21}Sn_2B_6$ with x=5 annealed at 810 °C for 50 hours and the simulation results.

The lattice constant obtained from the experimental data is 10.61 Å, and this value was used in the simulation. The occupational factor employed was 0.95 for Ni and 0.05 for Fe.

Figure 2 shows the Fe content in the τ phase, as measured by EPMA as a function of the lattice constants **a** measured for the τ phase. A sigmoidal dependence of $X_{Fe}/(X_{Fe}+X_{Ni})$ vs **a** is observed with a maximun Fe concentration of 0.28.

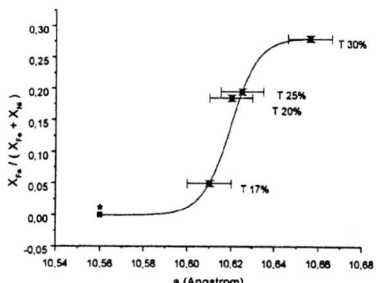

Fig.2: Fe content in the τ phase as a function of the lattice constants **a** measured . * τ phase of composition $Ni_{21}B_6Sn_2$.

Mössbauer Spectroscopy

Mössbauer spectra from the as cast samples are shawn in Figure 3. All of the spectra consist of four hyperfine interactions. The two with the smaller QS, belong to the τ phase. The other two belong to Ni_3Sn_2 with Fe in solid solution. The hyperfine parameters are reported in Table III.

Table III: Mössbauer Hyperfine parameters of τ phase and of (Ni, Fe)₃ Sn₂

SAMPLE	PHASE	IS_1 (mm/s) ±0.01	QS_1 (mm/s) ±0 01	$AREA_1$ %	IS_2 (mm/s) ±0.01	QS_2 (mm/s) ±0.01	$AREA_2$	area(Tau) area(Ni₃Sn₂)
T5%	τ	0.18	0.14	44.7	0.02	0.17	29.8	2.92
	Ni3Sn2	0.06	0.89	1.6	0.07	1.18	23.9	
T15%	τ	0.16	0.25	30.4	0.03	0.31	20.4	1.03
	Ni₃Sn₂	0.04	0.90	2.3	0.09	1.14	47.0	
T17%	τ	0.16	0.34	18.9	0.02	0.37	12.6	0.46
	Ni₃Sn₂	0.05	0.81	21.5	0.08	1.11	47.0	
T20%	τ	0.14	0.36	23.2	-0.02	0.46	15.6	0.63
	Ni₃Sn₂	0.05	0.92	46.5	0.09	1.09	14.7	
T25%	τ	0.14	0.35	19.1	-0.02	0.45	12.7	0.47
	Ni₃Sn₂	0.06	0.87	47.6	0.07	1.22	20.6	
T30%	τ	0.13	0.39	16.2	-0.02	0.43	10.9	0.37
	Ni₃Sn₂	0.05	0.83	39.5	0.07	1.08	33.5	

In spite that the τ phase has three probable sites for Fe the third site population is so small that it could not be observed. As the relative occupation of sites 1 and 2 in the τ cell is 3:2, under the supposition of equal probability, it is considered that the relative area A1/A2 will also be 3:2.

The QS from both τ sites increases and the IS decreases marginally as the overall Fe concentration increases to ≈20%. This is in agreement with the high Fe concentration, $X_{Fe}/(X_{Fe}+X_{Ni})$, up to 28 at%, measured by EPMA on the τ slabs.

The hyperfine parameters corresponding to both sites in Ni_3Sn_2 do not depend on the Fe concentration. This may be attributed to the low probability of finding Fe atoms as next neighbours of Fe atoms due to the low Fe concentration in this phase, $X_{Fe}/(X_{Fe}+X_{Ni})$ up to 7.5%.

Mössbauer spectra of samples T5% and T15% annealed at 810C for 50 hours show an increase in the area of τ at the expenses of Ni_3Sn_2. The presence of a Fe-Ni magnetic phase is observed in the T15% annealed sample. The hyperfine parameters corresponding to the Fe-Ni phase are IS=-0.06mm/s; QS=0.02mm/s and Bhf=28.7T.

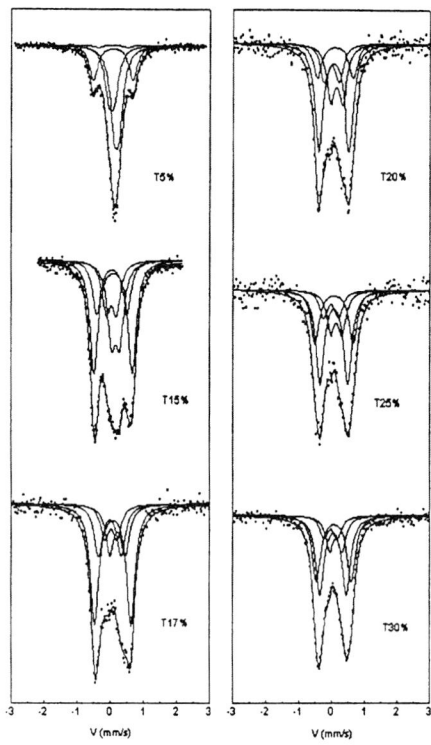

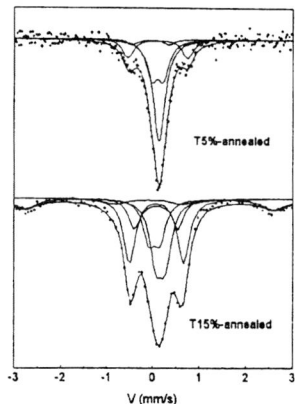

Fig. 4: Mössbauer spectra of samples T5% and T15% annealed at 810C for 50 hours.

Fig.3: Mössbauer spectra from as cast samples T5% to T30%

CONCLUSIONS

A quaternary τ phase exists in the Ni-Fe-B-Sn system with composition $(Fe_xNi_{100-x})_{21}B_6Sn_2$. Fe is dissolved in the Ni sites in the composition range $0\leq x \leq 28$. The hyperfine parameters, as well as the XRD pattern of the τ phase with x=5 are reported. When the lattice constant measured by means of XRD is correlated with the Fe relative concentration measured with EPMA a sigmoidal dependence is observed.

A second solid solution with composition $(Ni_{100-x}Fe_x)_3Sn_2$ ($x \leq 7.5$) was observed in this quaternary system. Its hyperfine parameters are also reported.

ACKNOWLEDGEMENTS

The authors thank F. Audebert for the CaRIne Crystallographic Software facility, and for the helpful discussions about the XRD analysis.

REFERENCES

[1] H. Z. Dokuzoguz, H. H. Stadelmaier, and L. H. Bowen, J. Less Comm. Met. 23 (1971) 245.
[2] L. H. Bowen, K. A. Taylor, H. Z. Dokuzoguz, and Stadelmaier, Proc.of the 7th Symposium on Mössbauer Effect Methodology, ed. I. J. Gruverman, vol. 7, (Plenum Press, New York, 1971) p.233.
[3] H.H. Stadelmaier, L.T. Jordan; Z. Metallkde 53 (1962) 719.
[4] H.H. Stadelmaier and C. B. Pollock; Z. Metallkde 60 (1969) 960.
[5] H.H. Stadelmaier, L.T. Jordan; Z. Metallkde 53 (1962) 719.
[6] M.Boudard, B. Arcondo, H.Sirkin; J. Mater. Sci., 26 (1991), 6517.
[7] B. Arcondo, M. Boudard and H. Sirkin, Hyp. Int. 67, (1991),677.
[8] V. Cremaschi, A. Ureña, J. Moya, F. Audebert, B.Arcondo y H. Sirkin, Proc. Anales de la Asociación Química Argentina. Vol. 83 N 5, (1995) 291.
[9] A. Ureña, V. Cremaschi, B. Arcondo, H. Sirkin; Hyp. Int. C (1997),101.
[10] K.P. Mizgalski, O. Inal, F. Yost, M. Karnowsky, J. Mater. Sci., 16 (1981), 3357.
[11] P. Villars, "Pearson's handbook of crystallographic data for intermetallic phases".(1991) 2nd Edition, 1586.

STUDY OF MAGNETIC AND STRUCTURAL PROPERTIES OF THE FeMnAlCCu ALLOYS

B. Cruz[1], G. A. Pérez Alcázar[1] and Y. Aguilar[2]
[1]*Departamento de Física, Universidad del Valle*
[2]*Dpto de Sólidos y Materiales, Universidad del Valle*
A.A. 25360 Cali, Colombia

In this work we report experimental analysis of magnetic and structural properties of the FeMnAlCCu alloys within the $Fe_xMn_{0.913-x}Al_{0.075}C_{0.01}Cu_{0.002}$ and $Fe_xMn_{0.911-x}Al_{0.075}C_{0.01}Cu_{0.004}$ series, with $0.50 \leq x \leq 0.80$.

The magnetic properties analysis was done by Mössbauer spectroscopy at room temperature, ac susceptibility and calorimetric techniques. It was found that for both series of alloys, the ones with a bigger content of Mn have an antiferromagnetic behavior typical of the fcc phase, which tends to be paramagnetic with decreasing Mn content. For the 16.3 % at and 16.1 % at of Mn alloys a paramagnetic and ferromagnetic phase coexistence was observed, while for the lower Mn content alloys it is the ferromagnetic phase which prevails, this ferromagnetic behavior is associated with the bcc phase. These phases were corroborated by X-ray diffraction and microstructural analysis. By ac susceptibility, characteristic curves were obtained, which show an antiferromagnetic behavior due to the different couplings between <001>,<110>,<111> spin directions, at the same time showing a temperature and composition dependence. By ac calorimetric measurements complete the evidence of the transition temperatures between the different spin directions detected by ac susceptibility.

It can be observed the rise of Cu concentration from 0.2 % at and 0.4 % at does not influence significantly the magnetic and structural properties of the studied alloys.

Introduction

Lately many efforts have been done in the study of FeMnAlC austenitic alloys, which under different thermical treatments undergo structural changes [1-3] which in turn influence notably their mechanical properties. Further studies about magnetic properties [4] of these alloys have shown how magnetic transitions, under cooling, can be classified according with its microstructure: (a) for completely austenitic γ-steels, the transition is from paramagnetic to antiferromagnetic, and (b) for α+γ-steels, the transition is from superparamagnetic to antiferromagnetic. In the theoretical [5] interpretation of experimental results [6] it is postulated that in these steels the Fe-Mn interaction is antiferromagnetic, Al acts as diluter and the Fe-Fe interaction is weakly ferromagnetic. From literature, one can find numerous studies of FeMnAlC alloys, but Cu addition has not been reported before in spite of the well known fact that the corrosion resistance of "weathering" steel is due basically to their lower Cu content. In this work the obtained results by studying the magnetic and structural properties of FeMnAlCCu alloys are presented. These studies were done by X-ray diffraction, Mössbauer spectroscopy, ac susceptibility and ac calorimetry on the series with $Fe_xMn_{0.913-x}Al_{0.075}C_{0.01}Cu_{0.002}$ and $Fe_xMn_{0.911-x}Al_{0.075}C_{0.01}Cu_{0.004}$ series, with $0.50 \leq x \leq 0.80$.

Experimental

Samples were obtained following Chakrabarty's method [7], which guarantees the disordered character of these samples. Samples structure was confirmed by X-ray diffraction and their magnetic properties were studied by Mössbauer spectroscopy, whose spectra were obtained by a conventional spectrometer in the constant acceleration mode at room temperature. These spectra were fitted using the Normos program [8]. Besides, ac susceptibility and calorimetry were used in a temperature range between 12-300 K.

Results and Discussion

In Figure 1 the Mössbauer spectra and hyperfine field distribution are showed in the case of 0.2 % at Cu alloys. It is observed on Figure 1(a) that the bigger Mn content alloys (41.3-26.3 % at) show a spectra made of wide line. It is clearly observed a proportional decreasing of the mean hyperfine field with the Mn content. This confirms the antiferromagnetic character induced by the Mn on these alloys, results that agree with other authors analysis in FeMn alloys [9-11] and FeMnAl alloys [6]. In the hyperfine field distributions, the most probable sites shift toward lower fields as the Mn content decreases. Through X-ray diffraction it was found that these alloys have fcc structure. In the Figure 1(b) the alloys fewer content of Mn are presented along with the spectra and field distributions. In the case of the 21.3 % at. Mn alloy, the fitting was done using two distributions, one for high fields (16-32 T) and the other for low fields (0-4 T). The fitting results show that the spectral area is composed of a 20.3 % coming from the higher fields and a 79.7 % from the lower fields. In the case of the 16.3 % at. Mn alloy, the fitting was done putting one distribution for the magnetic part (72.8 % of area) with a maximum of probability around a value H_{max} of 27.2 T which is characteristic of a ferromagnetic phase and with a paramagnetic site. Results from X-ray diffraction show the coexistence of austenitic (fcc) and ferritic (bcc) phase which agree with the ones obtained by Mössbauer spectroscopy. In the last alloy, with a 11.3 % at. content of Mn it is observed just a magnetic spectra with H_{max} equal to 29.0 T that is totally ferromagnetic due to the high content of Fe (80 % at.). This bcc related ferromagnetism is confirmed by a X-ray diffraction analysis [12].

The Mössbauer spectra and hyperfine field distributions corresponding to the 0.4 % at. Cu alloys (not shown) have a similar behavior to that obtained for 0.2 % at. Cu alloys (Fig.1). However, if the 21.1 and 21.3 % at. Mn spectra are compared, it can be seen that the spectra for the first alloy is composed only of one (fcc) singlet while that of the second one also shows a small percentage (20.3 % of area) of a ferromagnetic component (bcc). This can be justified by the antiferromagnetic behavior of Cu when in small quantities, so, having a higher content of this element

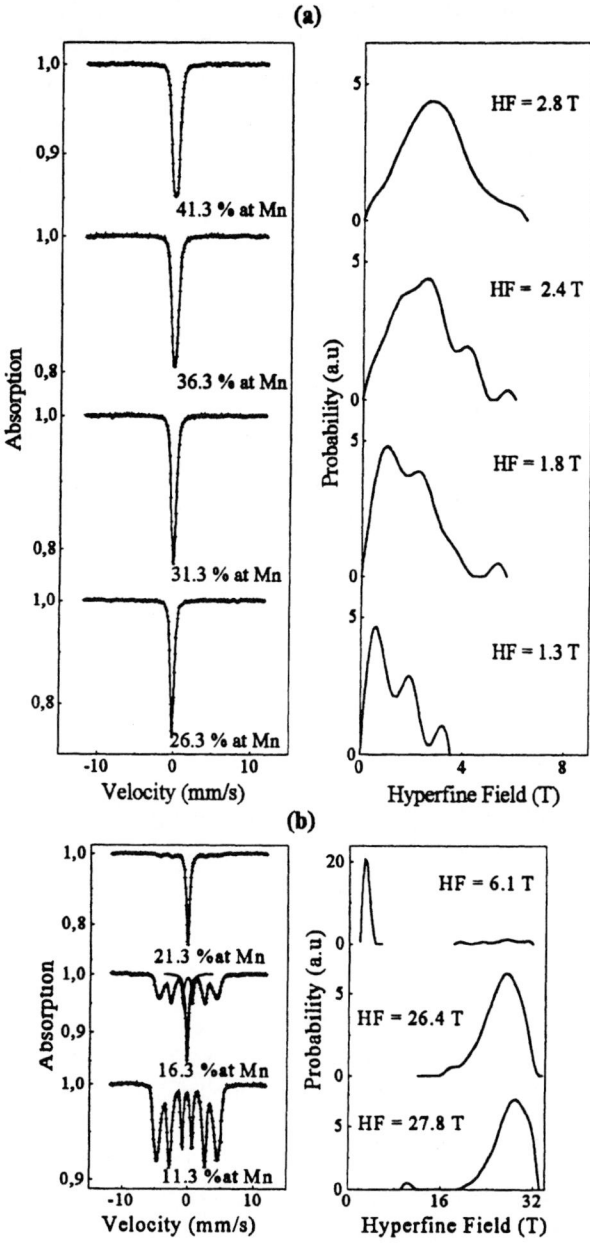

Figure 1. Mössbauer spectra and Hyperfine Field Distribution for 0.2 at % Cu alloys
(a) for the 41.3-26.3 at % Mn, (b) for the 21.3-11.3 at % Mn.

B. Cruz et al. / Study of magnetic and structural properties
of the FeMnAlCCu alloys

helps to settle the antiferromagnetic phase making the ferromagnetic one less probable. In the Table 1 the obtained Mössbauer parameters for the fitting of 0.2 and 0.4 % at. Cu alloys are summed up.

In Figure 2, a comparison of the mean hyperfine fields as functions of concentration (% at) of Fe between the studied alloys in the fcc phase is made. It can be observed a slightly higher mean hyperfine field for the 0.4% at Cu alloys. This agrees with the expected function of Cu as stabilizer agent.

Mn (% at)	IS	QS	Γ	H_{av} (T)	IS	QS	Mn (% at)	IS	QS	Γ	H_{av} (T)	IS	QS
41.3	-.26	.02	.30	2.8			41.1	-.22	.02	.30	3.1		
36.3	-.26	.02	.30	2.4			36.1	-.23	.01	.30	2.6		
31.3	-.26	.03	.30	1.8			31.1	-.22	.03	.30	2.1		
26.3	-.25	.06	.30	1.3			26.1	-.21	.06	.30	1.5		
21.3	-.19 -.26	0 .54	.35	6.1			21.1	-.22	.07	.35	1.1		
16.3	-.13	.01	.35	26.4	-.26	.46	16.1	-.14	.02	.35	26.9	-.23	.44
11.3	-.18	.01	.35	27.8			11.1	-.11	.01	.35	28.4		

Tabla 1. Mössbauer parameters obtained from the fitting of 0.2 and 0.4 % at Cu alloys (IS referred to α-Fe).
(Where not indicated the units are mm/s)

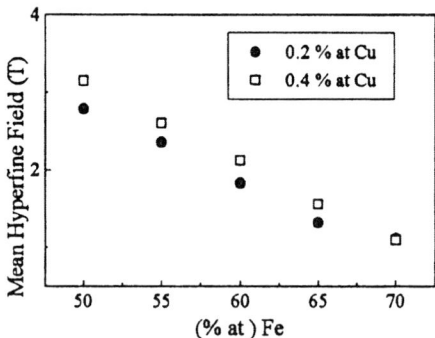

Figure 2. Mean Hyperfine Field (T) vs % at Fe for the two different Cu contents

In Fig. 3 the susceptibility curves as function of temperature for the 0.2% at Cu alloys with a Mn content between 41.3 and 16.3% at are shown. It can be observed a general behavior of the curves: they constantly rise and show two turning points which are more notorious in the curves with a higher Mn concentration. This can be explained taking into account the diluting function of the antiferromagnetic behavior of Fe in this phase. This turning points can be linked with a change in the antiferromagnetic coupling as the theoretical analysis of Ishikawa and Umebayashi suggests [13]. This analysis is based on the anisotropy energy minimization which leads to three different antiferromagnetic spin couplings with the same energy along the <001>, <110> and <111> directions, denoted in Fig. 3 as AF-I, AF-II and AF-III respectively.

Following the prior model, the first turning point can be explained as a transition of the antiferromagnetic coupling from the <001> to the <011> direction and the second point as a transition of the coupling from the <011> to the <111> direction. In the case of the 16.3% at Mn alloy, it can also be observed a constant rising but without clearly distinguished turning points. According to the Mössbauer analysis it was detected a ferromagnetic and paramagnetic phase coexistence, which indicates that the Neel temperature must be under room temperature. One possible explanation for this unobserved transition could be a shielding effect of the ferromagnetic phase with $T_c > T_{room}$.

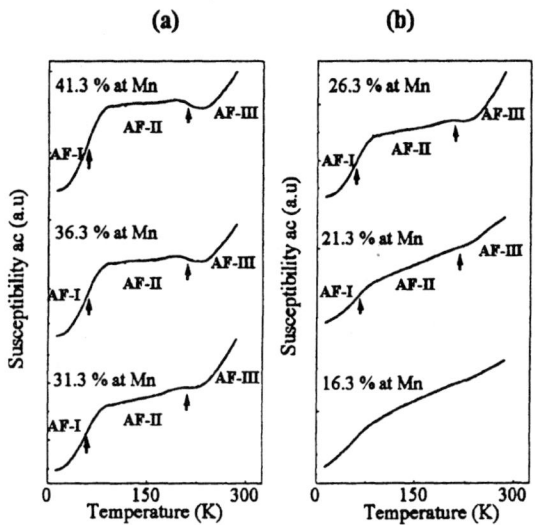

Figure 3.   ac susceptibility vs Temperature for the alloys with 0.2 % at Cu:
(a) from 41.3 to 31.3 % at Mn, (b) from 26.3 to 16.3 % at Mn.

With the purpose of demonstrating the appearance of the ac susceptibility observed transitions, ac calorimetry data was taken for the 36.3% at Mn alloy (0.2% at. Cu). Furthermore, the 36.1% at Mn alloy was also studied by this technique with the hope of elucidate the influence of a higher Cu content (0.4% at.) detected by Mössbauer spectroscopy.

In Fig. 4(a) the specific heat vs temperature of the 36.3% at Mn alloy is presented, where it can be observed a curvature change at T=61.8 K and an anomaly at T=216.5 K, this being into agreement with the ac susceptibility data for this alloy. In Fig 4(b) the specific heat vs temperature of the 36.1% at Mn alloy is presented. It can be observed a similar behavior, however, the anomalies appear at a slightly higher temperature. This temperature rising agrees with the Mössbauer spectroscopy analysis which showed that the alloys with a higher Cu content had a antiferromagnetic more stable phase and couplings.

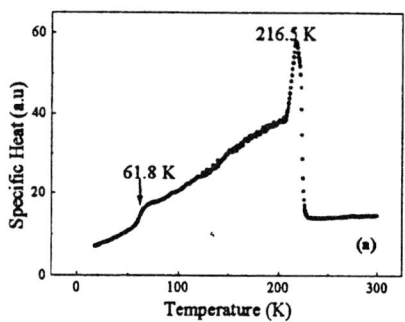 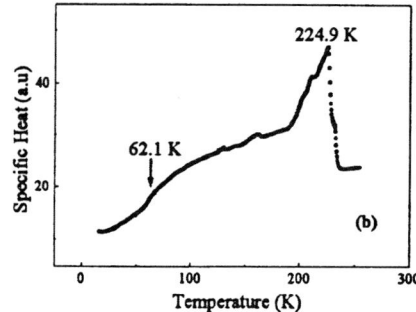

Figure 4.   Specific Heat vs temperature for the: (a) 36.3 % at Mn alloy with 0.2 % at
Cu, (b) 36.1 % at Mn alloy with 0.4 % at Cu.

Acknowledgments

The authors would like to thanks COLCIENCIAS and Universidad del Valle for
the financial support.

References

[1]K. H. Hwang and J. G. Byrne, Materials Science and Engineering, A 132
(1991) 161.

[2]H. Y. Chu, F. R. Chen and T. B. Wu, Scripta Mettallurgica et Materialia, 33(8)
(1995) 1269.

[3]K. H. Hwang, W. S. Yang, T. B. Wu, C. M. Wan, and J. G. Byrne, Acta Metall,
39 (1991) 825.

[4]S. U. Jen, Y. D. Yao, P. M. Huang, C. C. Lee, and S. C. Chang, J. Appl. Phys.
67(9) (1990) 4835.

[5]A. Osorio, Ligia E. Zamora, and G.A. Pérez Alcázar, Phys. Rev B 53, (1996)
8176.

[6]G. A. Pérez Alcázar, E. Galvao da Silva, and C. Paduani, Hyp. Int. 66 (1991)
221.

[7]Chakrabarti, Metall. Trans., 813, (1977) 121.

[8]R. A. Brand, Nucl. Instrum. Methods Phys. Res. B 28 (1987) 417.

[9]Y. Endoh and Y. Ishikawa, J. of the Phys. Soc. of Japan 30(6) (1971) 1614.

[10]Y. Ishikawa and Y. Endoh, J. of the Phys. Soc. of Japan 23(2) (1967) 205.

[11]N. S. McIntyre and D. G. Zetaruk, Anal. Chem. 49(11) (1977) 1521.

[12]G. A. Pérez Alcázar, J. A. Plascak, and E. Galvao da Silva, Phys. Rev. B 38,
(1988) 2816.

[13]H. Umebayashi and Y. Ishikawa, J. Phys. Soc. Japan 21 (1966) 1281.

TEMPERATURE DEPENDANCE OF SIGMA-PHASE FORMATION IN Fe-Cr-Mo-ALLOYS

F.B. Waanders[a], S.W. Vorster[a] and H. Pollak[b]

[a] School of Chemical and Minerals Engineering, PU for CHE, Potchefstroom, South Africa.
[b] Department of Physics, University of the Witwatersrand, Johannesburg, South Africa.

Fe-Cr-Mo alloys are used at high temperatures and under severe corrosive conditions but σ-phase precipitation in the 600-900°C temperature range adversely affects the mechanical properties of the alloy. Fe-Cr-alloys containing 2, 4 and 6% Mo were prepared and the transformation to the σ-phase was carried out by isothermally annealing the samples. Room temperature CEMS-spectra were recorded. Assuming precipitation kinetics to follow a Zener-Wert-Avrami type relation, activation energies were calculated and found to be 238 (±5) kJ/mole, 264 (±10) kJ/mole and 283 (±5) kJ/mole for the 2, 4 and 6% Mo-alloys respectively. Increasing the annealing temperature markedly accelerated the transformation rate and hardening of the samples.

1. Introduction

The demand for alloys that can be used at elevated temperatures and corrosive environments has been the driving force for investigations of the factors responsible for, amongst others, the formation of σ-phase, spinodal decomposition, embrittlement and other detrimental properties of the alloys. The understanding of precipitation of such phases and the effect on the alloys has been the subject of various research projects in the early parts of the twentieth century [1-4]. In the last fifty years the research has shifted to the high chromium steels, which have found a wide application in various branches of industry, especially those concerning high temperature erosion and wear [5-7]. The alloys suffer, however, from brittleness originating from the main component iron and chromium (Fe-Cr ≥90 %). The brittleness has two origins:

- the formation of the Cr-rich α'-phase that is the product of the spinodal decomposition that occurs if the material is annealed at T<510°C [8,9],
- the presence of the σ-phase that precipitates if the material undergoes a heat treatment at 510°C T 875°C [8,9].

In order to investigate the effect of temperature and composition on the kinetics of ageing, laboratory prepared Fe-Cr-Mo-alloys have been aged in the temperature range of 600°C T 800°C for various periods of time.

2. Experimental
2.1 Materials and ageing

The Fe-Cr-Mo-alloys used in this experiment were prepared in a vacuum arc furnace with a copper hearth and tungsten electrode by melting together appropriate amounts of high purity iron, chromium and molybdenum. To ensure homogeneity the samples were melted three times and finally quenched by contact with the water-cooled copper surface, which ensured cooling from the molten state to room temperature within half a minute or less. One side of the sample was ground to a roughness of 600μm, resulting in a flat 1cm^2 surface, ready for scanning electron microscopy and conversion electron Mössbauer spectroscopy (CEMS).

The samples produced in this investigation had compositions $Fe_{56.5}Cr_{41.6}Mo_{1.9}$, $Fe_{54.3}Cr_{42.1}Mo_{3.6}$ and $Fe_{52.8}Cr_{41.7}Mo_{5.5}$ respectively, determined by means of a scanning electron microscope (SEM) fitted with an energy dispersive X-ray spectrometer (EDAX). Transformation to the σ-phase was carried out by isothermally annealing the samples for various periods in an argon atmosphere at 600°C, 650°C, 700°C, 750°C and 800°C respectively. After an annealing cycle, the sample was heated to 950°C in an argon atmosphere for at least an hour and then quenched to ensure that the σ-phase was not present when the next cycle at another temperature commenced.

2.2 Mössbauer spectroscopy

After each annealing cycle the identification and evolution of the relative contribution of the σ-phase for each sample was studied by conversion electron Mössbauer spectroscopy CEMS) at 295 K. The Mössbauer source was 50 mCi ^{57}Co plated into a Rh foil. All spectra were obtained with the aid of a Halder Mössbauer spectrometer, capable of operating in conventional constant acceleration mode using a backscatter-type gas flow detector. A sample that contains both the σ-phase and ferrite will give rise to a paramagnetic singlet and a ferromagnetic sextet respectively. The amount of each constituent was determined from the areas under the relevant peaks.

2.3 Micro hardness measurements

Typical grain sizes of the samples were on average 150μm and it was thus possible to perform micro hardness measurements with a Leco M400 Vickers indenter under a load of 500 g. The measurements were repeated at least ten times and averages were taken.

3. Results and discussion

A typical Mössbauer spectrum for the as quenched, not annealed $Fe_{56.5}Cr_{41.6}Mo_{1.9}$-alloy is shown in figure 1(a). Before annealing, about 5% σ-phase was present in the as-quenched sample, possibly due to a small amount of σ-phase that formed during the initial production of the sample [10]. At an annealing temperature of 600°C the transformation of the sample to the σ-phase took about 40 hours, which reduced to less than 15 minutes at an annealing temperature of 800°C (see figure 1(b)).

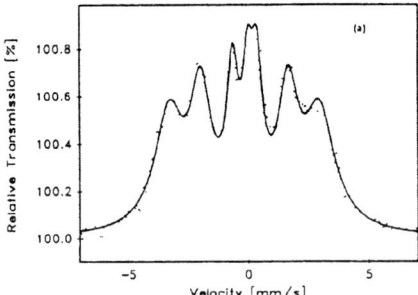 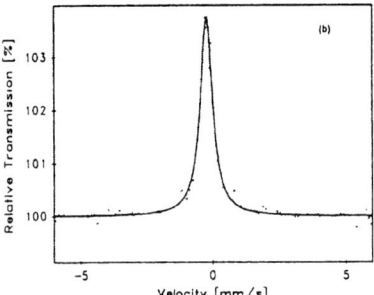

Figure 1. A Mössbauer spectrum of the $Fe_{56.5}Cr_{41.6}Mo_{1.9}$-alloy (a) without heat treatment and (b) the same alloy annealed for 15 min. at 800°C. A least-squares fitting to the data points was done and the envelope of the fit is shown in the figure.

Annealing the $Fe_{54.3}Cr_{42.1}Mo_{3.6}$-alloy at 600°C required 18 hours to transform, whilst at a temperature of 800°C it took only 12 minutes to transform to the σ-phase. In the case of the $Fe_{52.8}Cr_{41.7}Mo_{5.5}$-alloy transformation to the σ-phase at a temperature of 600°C took about 1 hour and less than 10 minutes at a temperature of 800°C. In a previous investigation [10], it was found that a $Fe_{50.7}Cr_{49.3}$-alloy completely transformed to the σ-phase within 72 hours at an annealing temperature of 800°C. These results indicate the influence of temperature and the addition of Mo to the rate at which the σ-phase forms in a Fe-Cr-alloy.

The ageing time t, required for attaining up to 50% transformation to the σ-phase will follow a Zener-Wert-Avrami-type relation, expressed by the following equation [11]:

$$\ln(1/t) = \ln A - (Q/RT) \qquad (1)$$

where Q is the activation energy, R the gas constant and T the temperature. A is a constant depending on the transformation conditions. From figure 2 it is clear that, at the lower annealing temperatures, after about 50% of the material has transformed, the rate decreases, possibly due to the formation of the χ- and R-phases [12]. These phases were however not observed in the Mössbauer spectra.

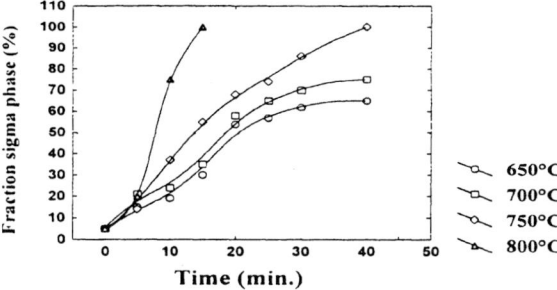

Figure 2. The fraction σ-phase that formed in the $Fe_{54.3}Cr_{42.1}Mo_{3.6}$-alloy in an annealing time of 50 minutes. A least squares fitting program was used to fit the lines.

From similar graphs for the different alloys studied the time t, for 25% and 50% transformation of the initial material to the σ-phase, was deduced. Little change in the slope of the lines indicate no change in decomposition and thus, according to equation 1 a plot of $\ln 1/t$ against 1/T would yield a straight line (see figure 3) from which the activation energies (Q) could be calculated.

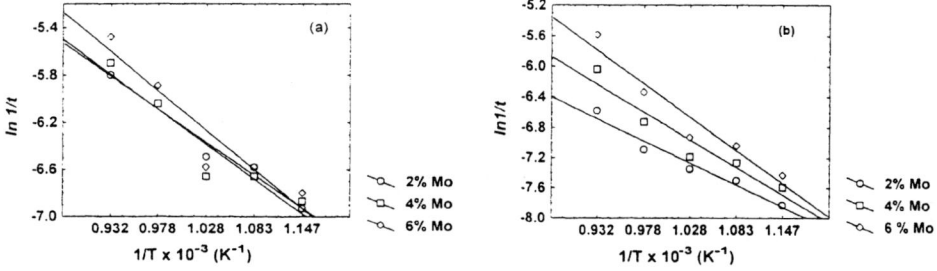

Figure 3. Plot of the rate ($\ln 1/t$) of formation of the σ-phase against 1/T for (a) 25% transformation and (b) 50% transformation. A linear regression was fitted to the data points to obtain the curves from which Q was calculated.

In table 1 the activation energies for the different alloys studied, are given, and the values obtained are consistent with values from literature [13-16]. In a previous investigation Waanders and Vorster [10] determined an activation energy Q = 106.2 kJ/mole for the $Fe_{54.3}Cr_{42.1}Mo_{3.6}$-alloy and concluded that the low value was possibly due to grain boundary diffusion. It however became evident in the present experiment that the low value was incorrect and the present results should be accepted. Brown and Smith [13] calculated the activation energies for model alloys from hardness measurements and found that the activation energy of Fe-Cr-Ni-alloys varied between 225 and 262 kJ/mole. These values correspond well with the activation energy of chemical diffusion in ferrite (230 kJ/mole) [14], with values calculated from the hardening behaviour of these alloys [15] and with the results presented in this investigation for the Fe-Cr-Mo-alloys. Ishikawa *et al.* [16] calculated an activation energy of 220(!10) kJ/mole for a Fe-28%Cr sample and 177(!3) kJ/mole for a Fe-28%Cr-5% Ni alloy which is lower than the values calculated in the present experiment for alloys containing Mo.

Table 1. Activation energies for the alloys studied

Alloy	Q 25% transformation kJ/mole	Q 50% transformation kJ/mole	Q Average KJ/mole
$Fe_{56.5}Cr_{41.6}Mo_{1.9}$	233	242	238!5
$Fe_{54.3}Cr_{42.1}Mo_{3.6}$	246	281	264!10
$Fe_{52.8}Cr_{40.5}Mo_{5.5}$	278	289	283!5

Micro hardness measurements [10,17] increase initially sharply and reach a maximum when the sample has been completely transformed to the σ-phase. Typical starting hardness values for all samples were 140 HV and after transformation took place the hardness increased to about 200 HV. Graphical representations can be found in references [10,17].

4. Conclusion

The σ-phase precipitation studied in this investigation follow a Zener-Wert-Avrami kinetics. The amount of Mo present in the alloy and the ageing temperature both has a marked influence on the rate of the σ-phase formation. From a graph of the precipitation rate against temperature, the activation energy for each alloy was calculated.

References

[1] P.D. Merica, Trans. Metal. Soc. of AIME, 99, (1932), 13.
[2] R. Becker, Z. Metallk., 29, (1937), 245.
[3] U. Dehlinger, Z. Metallk., 29, (1937), 401.
[4] A.J. Lena, Metal Progress, July, (1954), 86.
[5] M. Kato, Acta Mettalurgica, 29, (1981), 79.
[6] S. Bonnet et al., Materials Science and Technology, March, 6, (1990), 221.
[7] J-P Massoud *et al.*, Sixth International Symposium on Environmental Degradation of Materials in Nuclear Power Systems-Water Reactors Edited by R.E. Gold and E.P. Simonen, The Minerals, Metals and Materials Society, (1993), 399.
[8] P. Marshall, *Austenitic Stainless Steels*, Elseviers Applied Science Publishers, London, UK, (1984), 184.
[9] S.M. Dubiel and G. Inden, Z. Metallkunde, 78 (1987), 544.
[10] F.B. Waanders and S.W. Vorster, Hyp. Int. 112, No1-4, (1998), 139.
[11] J. Burke, *The kinetics of Phase Transformations in Metals*, Pergamon Press, (1965), 55.
[12] H.D. Solomon and T.M. Devine in *Duplex Stainless Steels* edited by R.A. Lula, American Society of Metals, Ohio, 1983.
[13] J.E. Brown and G.D.W. Smith, Surface Science, 246 (1991), 285.
[14]P.J. Grobner, Trans. Am. Inst. Min. Metall. Pet. Eng. 4a, (1973), 251.
[15]J.E. Brown *et al.*, Fatigue, Degradation and Fracture, Published by A.S.M.E., New York, Edited by W.H. Bamford *et al* PVP vol. 195, MPC vol. 30, 1990, 175.
[16] Ishikawa *et al.* Materials Transactions, JIM, 36 (1), (1995), 16.
[17] F.B. Waanders, H. Pollak and S.W. Vorster, To be published in Hyperfine Interactions

MÖSSBAUER SPECTROSCOPY AND EXAFS STUDY OF NANOSTRUCTURED NiZn- FERRITE

A. S. Albuquerque[1], J.D. Ardisson[1], W. A. A. Macedo[1] and M. C. M. Alves[2]

[1]*Laboratório de Física Aplicada, Centro de Desenvolvimento da Tecnologia Nuclear - CDTN*
C. P. 941, 30123-970 Belo Horizonte, Brazil. E-mail:asa@urano.cdtn.br
[2]*Laboratório Nacional de Luz Síncrotron - LNLS, Campinas, Brazil.*

The structure and the hyperfine properties of nanostructured powder of $Ni_{0.5}Zn_{0.5}Fe_2O_4$ were investigated by extended X-ray absorption fine structure spectroscopy (EXAFS), X-ray diffraction and Mössbauer spectroscopy. Samples of high purity and high homogeneity were obtained by coprecipitation method, followed by annealing between 300 and 1350° C. EXAFS was applied to follow Ni, Zn and Fe cations in the initial crystallization on the NiZn- ferrite. The mean diameter of the ferrite particles, in the nanometer range, was determined by XRD. EXAFS results show that the short range order of the samples changes with increasing calcination temperature. The mobility of Zn atoms is lower than Fe and Ni in the studied system. The complete structural order was achieved after heat treatment at 1350° C. Mössbauer results show the high purity and 1:1 Ni to Zn stoichiometry of the obtained ferrimagnetic $Ni_{0.5}Zn_{0.5}Fe_2O_4$ samples and the superparamagnetic behaviour of all nanosized samples.

The properties of nanostructured magnetic materials are highly affected by the microstructure that depends on the synthesis processes. While conventional methods to obtain ceramic materials involve high temperatures, chemical routes like coprecipitation, sol-gel and hydrothermal processing can provide convenient non-conventional ways to prepare ultrafine powders of high purity and homogeneity by using relatively low temperatures [1]. In this work, we have investigated the non-conventional synthesis of nanostructured ferrimagnetic NiZn-ferrite, $Ni_{0.5}Zn_{0.5}Fe_2O_4$ [2] prepared by coprecipitation, and investigated the structural and the hyperfine properties of the obtained samples.

The samples were synthesized by using Fe, Ni and Zn nitrates as precursor, which were dissolved in deionized water in the required mole proportion. NaOH (2.5 M) was used as precipitating agent and the obtained initial powder was submitted to treatment at temperatures between 300° C and 1350° C under ordinary atmosphere, for 2 hours (Table I) [3]. The chemical composition of the samples was checked by X-ray fluorescence (EDX) and the characterization was done by X-ray diffraction (XRD), extended X-ray absorption fine structure (EXAFS) and Mössbauer spectroscopy (MS). XRD was performed employing Cu-Kα radiation. EDX composition measurements were obtained after calibration with standards specimens prepared from Ni, Zn and Fe oxides of high purity. Mössbauer spectra were obtained in transmission mode, by using a standard Co (Rh) source.

EXAFS spectra at the Ni (8333 eV), Zn (9569 eV) and Fe (7111 eV) absorption K-edges were obtained of the XAS beamline at the synchrotron storage ring at Campinas (LNLS), running at a typical current on 80 mA. The energy of the photons was selected and scanned with a Si (111) channel cut double crystals monochromator with an energy resolution of 2 eV at Ni, Zn and Fe K-edges. The EXAFS spectra were analysed by using the standard methods [4], that involves a background subtraction, determination of absorption edge, normalization, simulation of atomic absorption, extraction of the EXAFS signal and obtaition, by Fourier transform, of the contributions from the different neighbour shells. By an inverse Fourier transform into K space, the EXAFS oscillations corresponding to the different neighbour shell were obtained. The structural parameters, number of neighbours (N), atomic distances (R), and Debye-Waller coefficient (σ) were taken from last square fitting using experimental phase and amplitude functions deduced from a model compound. Spectra collected from standards of Fe_2O_3, NiO, ZnO, NiZn-ferrite, Ni-ferrite and Zn- ferrite were also analysed and compared to our samples.

Figure 1 shows X-ray diffraction patterns of the NiZn-ferrite samples. The powder, as obtained, and the samples heat treated at 300° C show poor crystallization, with less defined diffraction lines, and we can observe the crystallinity evolution with increasing annealing temperature. Samples heat treated up to 400° C exhibit peaks attributable to NiZn-ferrite and lattice parameter was 8.39 (1) Å, as calculated by extrapolation, in according to Klug [5]. The average diameters of the particles (D), as obtained from XRD [5] are indicated in Table I.

Table I- Annealing temperature (T) and the particle mean diameter (D ±3 nm) of the NiZn- ferrite samples.

T (°C)	as obtained	300	400	600	700	800	900	1350
D (nm)	< 6	9	20	30	39	90	120	> 150

Figure 2 shows the EXAFS Fourier transform (FT) of the Fe, Ni and Zn K-edge EXAFS for our NiZn ferrite samples and for the measured standards. The different FT peaks have been identified to correspond to the coordination of these elements with oxygen and with the ions in the characteristic tetrahedral (A) and octahedral [B] sites of the spinel configuration of the NiZn-ferrite [2], represented here by -(A) and -[B]. For the quantitative analysis of the spectra from the Ni and Zn K-edge we have used phase and amplitude extracted from a well characterized NiZn-ferrite sample annealed at 1350° C, with parameters in accordance with other authors [6,7]. Since the corrections for electron phase shifts have not been incorporated, the FT spectra do not represent true radial atomic distance.

For the Ni K-edge FTs (Fig.2b), the first peaks refer to Ni-O, the second peaks refer to Ni-[B], and third peaks were identified as Ni-(A). We can note, from the increasing peaks amplitude, specially for the second and third peaks, the better crystallization of the structure with increasing temperature, as expected. The inverse FT analysis of the first peak, at R= 2.1 Å, indicates the octahedral coordination of the Ni by the oxygen (N= 6), for all the samples. As shown in Table II, for the second and third coordination shell it was observed that the sample without heat treatment presents structural parameters N and R in disagreement with the standard values. The evolution of the structural order with heat treatment is indicated the clear decrease of the Debye-Waller factor with increasing temperature.

For the Zn K-edge (Fig. 2-c), the first peaks correspond to Zn-O, the second peaks correspond to the Zn-[B] and Zn-O, and the third peaks refers to Zn-(A). The inverse FT analyses of the first peak indicates the tetrahedral coordination of the Zn by the oxygen (N=4). The structural parameters obtained from second and third coordination shells (Table II) indicated that

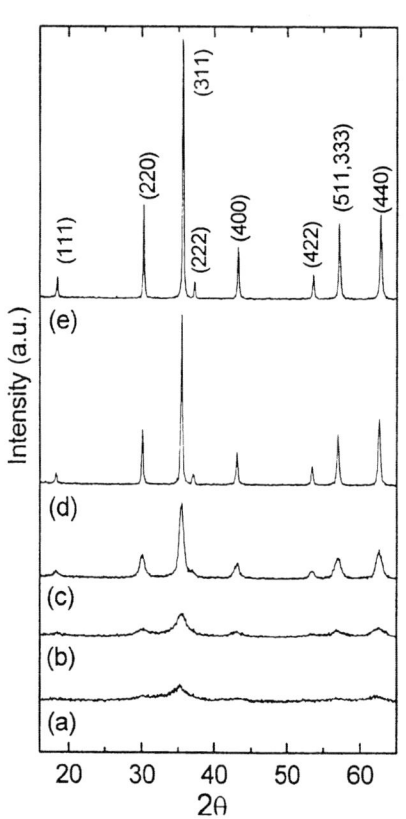

Figure 1- X-ray diffraction patterns of the NiZn-ferrite samples (a) as obtained and annealing at (b) 300°C, (c) 400°C, (d) 700°C and (e) 1350°C.

Table 2- Structural parameters obtained for the NiZn-ferrite samples, in the Ni and Zn K- edge.

T_{cal} (°C)		Ni K-edge					Zn K-edge				
		1350	700	400	300	as obtained	1350	700	400	300	as obtained
1st shell	N	6.0	6.3	5.7	5.7	5.7	4.0	4.0	3.8	3.7	3.6
	R (Å)	2.1	2.1	2.1	2.1	2.1	1.90	1.89	1.89	1.88	1.89
	σ^2	0	0.0012	0.0013	0.0012	0.0014	0	0.0001	0.0004	0.0005	0.0007
2nd shell	N	6.0	6.0	5.8	5.6	2.8	12.0	11.5	11.0	9.5	6.7
	R (Å)	2.96	2.96	2.96	2.97	3.01	3.50	3.49	3.49	3.49	3.49
	σ^2	0	0.0002	0.0006	0.0024	0.0045	0	0.0013	0.0043	0.0105	0.0158
3nd shell	N	6.0	5.80	5.5	5.0	5.1	4.0	4.0	4.0	2.7	1.4
	R (Å)	3. 20	3.20	3.21	3.21	2.8	3.80	3.81	3.82	3.83	3.85
	σ^2	0	0.0001	0.0011	0.0020	0.0027	0	0.0002	0.0045	0.0073	0.0087

the expected Zn distribution in the NiZn-ferrite occurred even prior the heat treatment, but with high degree of disorder.

The analysis of the Fe K-edge spectra (Fig. 2-a) are more complex since Fe atoms are present at A and B sites. The FT peaks centred near 1.5 Å correspond to $Fe_{(A)}$-O and $Fe_{[B]}$-O, the peaks near 2.6 Å correspond to $Fe_{[B]}$-[B] and the peaks near 3.2 Å include contributions from $Fe_{(A)}$-(A), $Fe_{(A)}$-O, $Fe_{[B]}$-(A) and $Fe_{[B]}$-O. This complexity arises from multiple scattering effects, fact that do not allow the fit of these EXAFS spectra by using experimental phase and amplitude. Then, for Fe K- edge we have used FEFF 6 multiple scattering code [8], in order to obtain theoretical phase and amplitude from Fe_2O, taken as reference. For the peak centred near 1.5 Å, the simulation considered contributions of two Fe-O distances. For the Fe atoms in our NiZn-ferrite samples, these procedure resulted in a number of oxygen neighbours varying from 4 to 5, and Fe-O distances differing by about 0.18 Å for A and B sites. The quantitative analysis of the second and third coordination is more complex and it will be considered for publication elsewhere.

Figure 3 shows Mössbauer spectra at room temperature of the nanosized powder samples, as obtained, after coprecipitation, and after heat treatment. Samples annealed at T < 400° C reveal strong superparamagnetic behaviour characterized by the collapse of six lines spectra. The superparamagnetism decreases

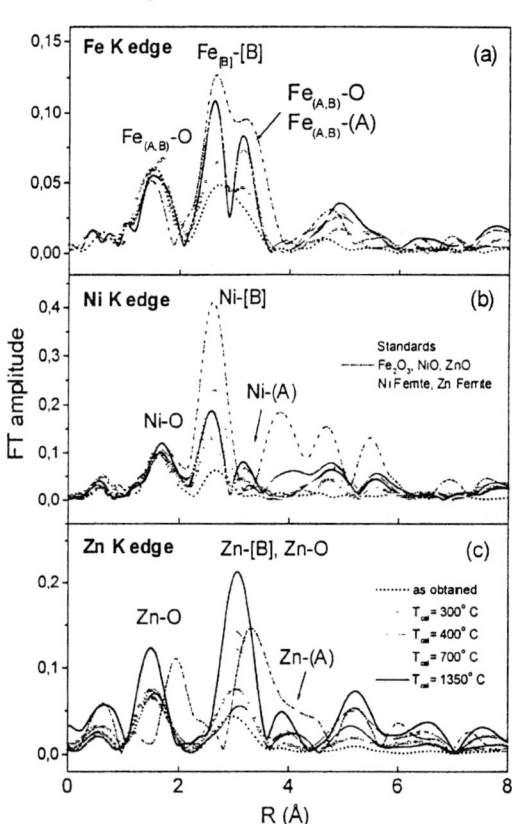

Figure 2- EXAFS Fourier transform of the (a) Fe, (b) Ni and (c) Zn K-edges, obtained from NiZn-ferrite samples and from the standards Ni ferrite, Zn ferrite, NiO, ZnO and Fe_2O_3.

with increasing particles mean diameter and, starting from D = 30 nm (T = 600° C), the magnetic hyperfine structure becomes resolved, nevertheless with mean hyperfine field (B_{hf}) lower than the expected one for bulk ferrite [1]. The resolved spectra were fitted to one sextet with linewidth near 0.6 mm/s attributed to Fe^{3+} at A sites, and one magnetic hyperfine field distribution relative to Fe^{3+} in B sites, as shown in Figure 3-d e 3-e. The use of a B_{HF} distribution to fit the sextet corresponding to Fe at B sites can be explained by the presence of different magnetic neighbours affecting the iron atoms distinctly [9].

For the ferrimagnetic Mössbauer spectra (fig. 3-d and 3-e), the obtained isomer shifts (IS) are IS(A) = 0.34 (3) mm/s and IS[B] = 0.38 (3) mm/s (relative to α-Fe), quadrupole splitting (QS) near zero for the two sites, and magnetic hyperfine fields (B_{hf}) varying from 43.5 to 46.9 for A sites and from 40.0 to 45.0 for B sites. These B_{hf} values did not correspond to well crystallized $Ni_{0.5}Zn_{0.5}Fe_2O_4$ [1], indicating some superparamagnetic relaxation even for the resolved spectra. The obtained area ratio between the B and A sites was close to 3, in coherence with the cation distribution of a stoichiometric $Ni_{0.5}Zn_{0.5}Fe_2O_4$, where Ni^{2+} ions are expected to occupy only octahedral sites and Zn^{2+} ions are expected to occupy only tetrahedral sites [10].

In conclusion, we have investigated the formation of stoichiometric nanocrystalline NiZn-ferrite by coprecipitation followed by heat treatment at relatively low temperatures. EXAFS measurements were applied to follow the initial crystallization of the samples and these results indicated that the formation of NiZn ferrite starts even before heat treatment, although with some structural disorder. Mössbauer spectroscopy showed that, at room temperature, ferrite powders with particles mean diameter smaller than 30 nm the exhibit strong superparamagnetic relaxation. Bigger particles showed ferrimagnetic spectra with reduced magnetic hyperfine fields.

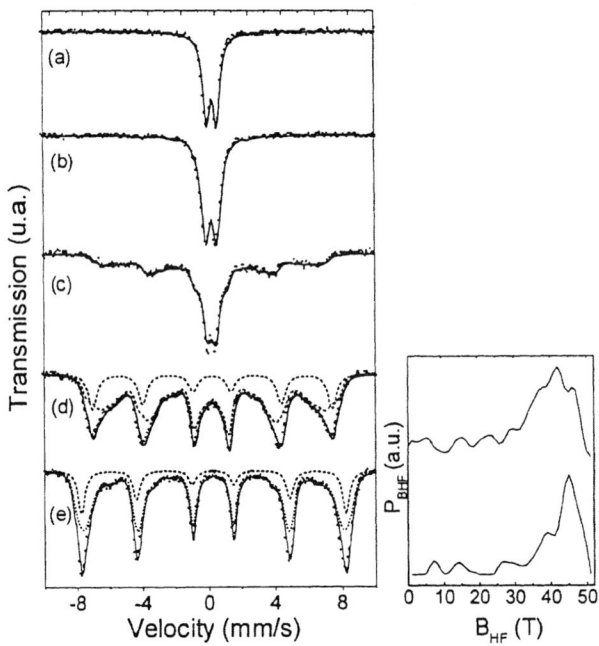

Figure 3- Mössbauer spectra, at room temperature of the NiZn-ferrite samples (a) as obtained and heat treated at (b) 300° C, (c) 400° C, (d) 700° C and (e) 1350° C, and the magnetic field distributions referring to the (d) and (e) spectra.

References

[1] D.L.L. Pelecky and R.D. Rieke, Chem. Mater. 8 (1996) 1770.

[2] J.M. Daniels and A. Rosencwaig, Canadian J. of Phys. 48 (1970) 381.

[3] A.S. Albuquerque, J.D. Ardisson, N.D.S. Mohallem and W.A.A. Macedo, Hyp. Int. (C) 3 (1998) 256.

[4] D.E. Sayers, E.A. Stern and F.W. Lytle, Phys. Rev. Lett. 27 (1971) 1204.

[5] H.P. Klug and L.E.Alexander, in X-ray Diffraction Procedures for Polycrystalline and Amorphous Materials, John Wiley & Sons (Wiley-Interscience, New York, 1974) p.594 .

[6] V.G. Harris, N.C. Koon, C.M. Williams, Q. Zhang and M. Abe, J. Appl. Phys. 79 (1996) 4561.

[7] A. Claasen, Proc. Phys. 38 (1925-1926) 482.

[8] J.J. Rehr, R.C. Albers, S.S. Zabinsky, Phys. Rev. Lett 69 (1992) 3397.

[9] T.A. Dooling and D.C. Cook, J. Appl. Phys. 69 (1991) 5352.

[10] C.L. Chien, Annu. Rev. Mater. Sci. 25 (1995) 129.

STRUCTURAL AND MÖSSBAUER STUDIES OF MILLED Al_xFe_{100-x} OVER THE ENTIRE COMPOSITION RANGE

M. Meyer, L.A. Mendoza-Zélis, F.H. Sánchez

Departamento de Física, Universidad Nacional de La Plata, CC 67, 1900 La Plata, Argentina

We present here some results obtained by milling the Al-Fe system in the entire composition range. The resulting phases were analyzed by X ray diffraction and Mössbauer spectroscopy. For aluminum contents up to 60 at% a bcc random solid solution (A2 structure) was obtained, with some B2 ordering in the case of Al richer samples. For the composition $Al_{70}Fe_{30}$ the compound Al_5Fe_2 was formed coexisting with the bcc phase, whereas for $Al_{80}Fe_{20}$ the main formed phase was $Al_{13}Fe_4$. Finally for $Al_{90}Fe_{10}$ some pure Al persists coexisting with the compound $Al_{13}Fe_4$.

Introduction

The structure and magnetic properties of Al_xFe_{100-x} alloys have been intensively studied in the past few decades. [1,2,3]. Their production is more difficult for compositions with x>50 due to the complexity of the Al-Fe phase diagram in this region and to the very low solubility of Fe in Al. Different fabrication methods (e.g. coevaporation, liquid quenching, ion implantation or irradiation) were used that froze metastable states avoiding the inhomogenization via compound formation and segregation. On the other hand, mechanical milling has been widely used to produce this kind of metastable states. Starting from the elemental powders a steady state is reached which seems to depend not only on the nominal composition but also on milling parameters, like the so called milling intensity [4].

We have previously studied the evolution of Al_xFe_{100-x} powder mixtures under milling for x =80 and 85 [5] finding a final product which consists of a mixture of Al and some AlFe compounds. We have also studied the effect of milling on crystalline AlFe compounds with B2 structure, obtaining different steady values of the long range order parameter [6,7]. In this work, we report on an attempt to make up Al_xFe_{100-x} alloys, across the entire composition range, by mechanical alloying the elemental powders. The final products obtained under these milling conditions were studied by X ray diffraction (XRD) and Mössbauer Effect (ME) spectroscopy, allowing their structural characterization and the determination of the local magnetic field at the Fe sites.

Experimental

Al_xFe_{100-x} samples, with x ranging from 10 to 90 in steps of 10, were prepared by mixing the elemental pure Fe (Fluka 0.995 -325 mesh) and Al (Johnson-Matthey 0.995 -325 mesh) powders in adequate proportion to obtain the desired composition. Samples of 1 g each, together with two chrome steel balls (Φ = 12 mm), were sealed into chrome steel cylindrical vials (10 cm^3) under Ar atmosphere, after several Ar purges. They were then mechanically ground in a Retsch MM2000 horizontal vibratory mill operating at 30 Hz during time intervals ranging from 20 to 80 hours in order to reach the steady state for each composition. The estimated milling intensity under these conditions was 450 m/s^2.

After milling, the samples were characterized by XRD and ME spectroscopy. XRD patterns were obtained in reflection geometry using a Philips PW1710 diffractometer with Cu Kα radiation. ME experiments were carried out with a conventional spectrometer under transmission geometry, employing a ^{57}CoRh radioactive source. All these measurements were performed at room temperature.

Results

The X ray diffractograms are shown in figure 1, where it can be seen that for Al contents up to 60 at%, the milling produced the formation of a bcc solid solution (A2 phase). For x = 50 an 60 there is an insinuation of B2 superstructure formation, evidenced by the presence of the (100) reflection, indicated in the figure. The shift in the peaks position of the A2-bcc phase towards lower angles, indicates the increase of the lattice parameter that can be seen in figure 2.

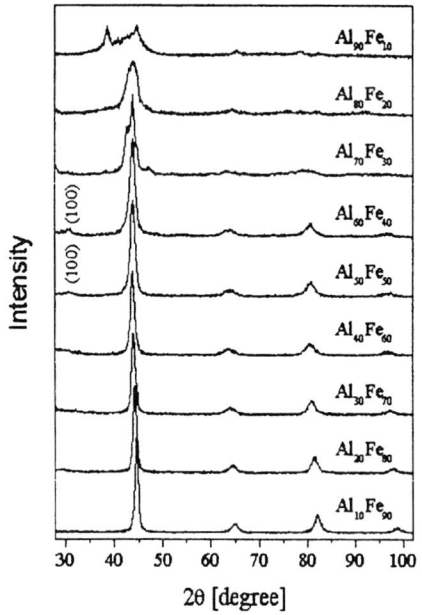

Figure 1. *X ray diffractograms for all the studied samples*

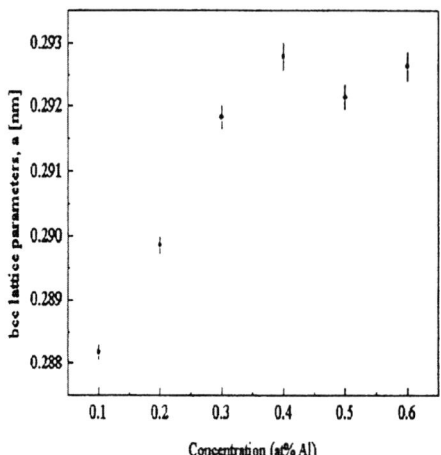

Figure 2. *Al content dependence of lattice parameter for bcc solid solutions.*

On the other hand, for higher Al contents and depending on composition, the reflections corresponding to some of the Al-Fe compounds that appear in the phase diagram can be observed. In the case of the Al$_{90}$Fe$_{10}$ sample, the XRD diffractogram shows the aluminum reflections superimposed to those presumably corresponding to $Al_{13}Fe_4$ compound.

The Mössbauer spectra for all the studied samples are shown in figures 3 and 4. A net dependence of the spectrum shape with composition may be observed: for $x \geq 50$ the spectra are no magnetic, whereas for the Fe rich samples a broad magnetic sextet is found. To analyze these spectra (either magnetic or not), several kinds of fits were attempted using standard least squares routines and including broadened Lorentzians, Voigt profile lines or model-independent hyperfine magnetic fields H or quadrupole splitting (Δ_{QS}) distributions [8]. Good fits were obtained with model independent distributions in the case of magnetic spectra. For the non-magnetic spectra the same kind of fit can be applied (with a Δ_{QS} distribution), but for those samples with $x = 70$ and 80 a two component model, taking into account the phases revealed by the diffractograms, increases the fit quality.

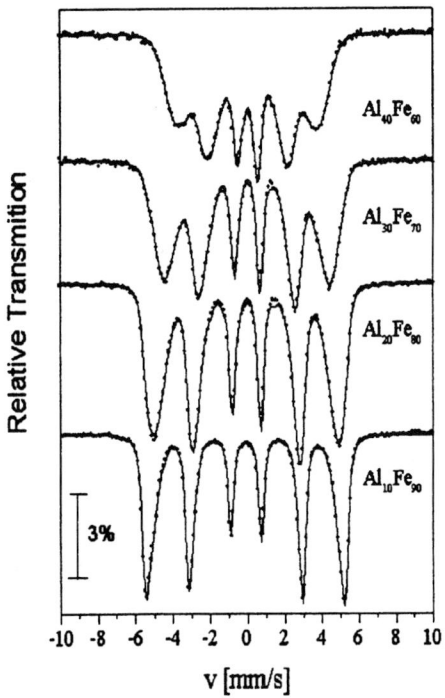

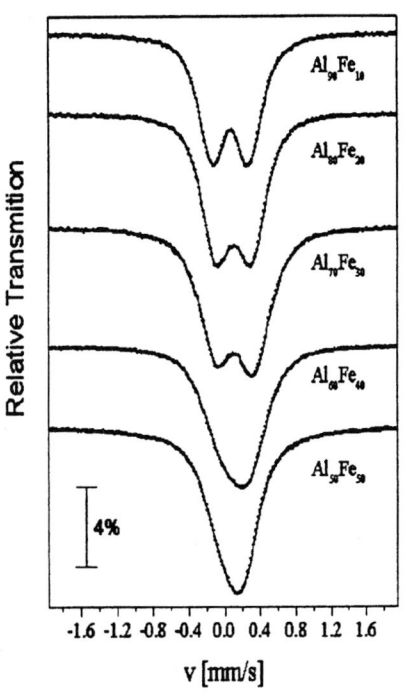

Figure 3. *ME spectra (high velocity) for Fe rich samples*

Figure 4. *ME spectra (low velocity) for Fe poor samples*

Discussion

In view of the results obtained with both techniques, it can be established that in the composition range from pure iron to 60 at% aluminum the milling produces the formation of a bcc solid solution. The XRD results indicate that some ordering takes place for $Al_{50}Fe_{50}$ and $Al_{60}Fe_{40}$ as evidenced by the apparition of the (100) peak that corresponds to an ordered B2 type structure.

The sample of composition $Al_{70}Fe_{30}$ can be understood in terms of the coexistence of bcc solid solution and the Al_5Fe_2 compound. On the other hand, the $Al_{80}Fe_{20}$ sample may be interpreted in terms of the coexistence of Al_5Fe_2 and $Al_{13}Fe_4$ compounds, although a third minority phase with bcc structure can not be disregarded. Finally, in the case of the $Al_{90}Fe_{10}$ sample, some unreacted aluminum remains together with $Al_{13}Fe_4$ compound.

Similar results were recently reported by Eelman et al. [9] who found the formation of a bcc solid solution up to 60 at %Al, with no indication of B2 ordering . Furthermore, Enzo et al [10] reported the formation of a bcc solid solution up to 75 at %Al, when using a milling device of higher intensity.

From the fits of the Mössbauer spectra for x<50 with a distribution of sextets an average value of the hyperfine magnetic field H was obtained and displayed in figure 5 as a function of composition. The observed behaviour confirms previous results on samples made by other techniques showing a reduction in hyperfine field with x and a clear drop to zero when x approaches 50. This reduction may be explained by the increasing average number of Al neighbours of Fe atoms.

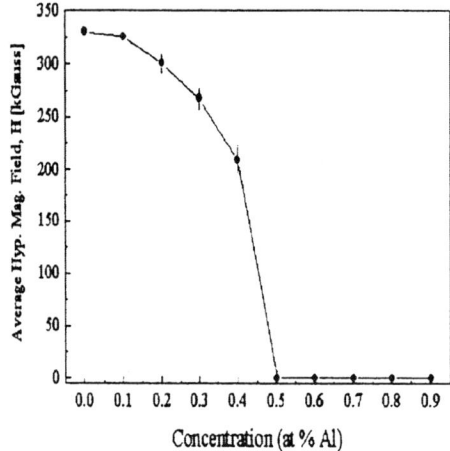

Figure 5. *Dependence of magnetic hyperfine field with composition*

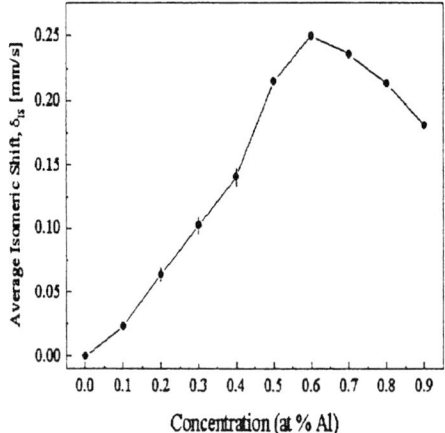

Figure 6. *Dependence of isomeric shift with composition*

Also the average value of the isomeric shift (δ_{IS}) may be displayed as a function of composition (see figure 6) for all the studied range. The general trend is similar to that observed in vapor quenched samples [1] and has been explained in terms of electronic properties[11]. The increase of δ_{IS} with x is caused by a charge transfer that occurs when atoms with different electronegativities are combined together. Also, the requirement that the electron density at the boundary of dissimilar atomic cells should be continuous forces atoms to redistribute their intra-atomic charges. This interpretation has been developed in the Miedema-van der Woude model [12]. On the other hand, the reduction of δ_{IS} for higher values of x, can be understood in terms of the hybridization between the Fe 3d-band and the Al sp-band. This will normally increase the width of d band, reduce the d character and increase the sp character in the band. This will not only suppress the shielding action of the d electrons to the s electrons but also increase the s-electron density at the nucleus, producing a decrease of the δ_{IS}.

Conclusion

Room temperature milling of elemental powders of Al and Fe produces a bcc Al_xFe_{100-x} solid solution, with a solubility range extended up to 60 at% of aluminum. For larger aluminum contents disordered equilibrium compounds are obtained. It is worth to notice that Al_5Fe_2 is the first compound to be formed when there is an Al excess. This may be related with the fact that Al_5Fe_2 is also the first compound to precipitate during the thermal mixing of Al/Fe multilayers [13]. The transformation o-Al_5Fe_2 +Fe $\leftrightarrow$ bcc-AlFe under milling merits further investigation.

Finally, comparing our results with those obtained by Enzo el al. [10] the idea that the final state depends on the milling intensity is confirmed: higher metastability for higher milling intensity.

References

[1] J.H. Hsu and C.L. Chien, Hyp. Int. **69** (1991) 451.
[2] G.P. Huffman and R.M. Fisher, J. Appl. Phys. **38** (1967) 735
[3] S. Nasu, U. Gonser and R.S. Preston, J. Physique **41** (1980) C1-385
[4] Y. Chen, M. Bibole, R. Le Hazif and G. Martin, Phys. Rev **B48** (1993) 14
[5] M. Meyer, L. Mendoza Zélis and F.H. Sánchez, Mat. Sci. Forum **225-227** (1996) 441
[6] M.T. Clavaguera-Mora, J, Zhu, M. Meyer, L. Mendoza-Zélis, F.H. Sánchez and N. Clavaguera, Mat. Sci. Forum **235-238** (1997) 541
[7] M. Meyer, L. Mendoza Zélis and F.H. Sánchez, Hyp. Int. (C) **2** (1997) 213
[8] R.A. Brandt, Nuc. Instr. Methods **B28** (1987) 398
[9] D.E. Eelman, J. R. Dahn, G.R. Mackay and R.A. Dunlap, J. Alloys Compounds **266** (1998) 234
[10] S. Enzo, G. Mulas and R. Frattini, Mat. Sci. Forum, **269-272**, (1998) 385
[11] C.L. Chien, G. Xiao ans S.H. Liou, Hyp. Int. **27** (1986) 373
[12] A. R. Miedema and F. van der Woude, Physica **100B** (1980) 145
[13] M. Meyer, L. Mendoza Zélis, F.H. Sánchez and A. Traverse, Hyp. Int. **83** (1994) 327

EVOLUTION OF THE SHORT RANGE ORDER PARAMETERS FOR THE TRANSFORMATION A2⇒DO$_3$ IN Fe$_{75}$Ge$_{25}$

A.F.. Cabrera F. H. Sánchez and L.A. Mendoza Zélis

Departamento de Física, FCE, Universidad Nacional de La Plata, C.C. 67, 1900 La Plata

In this contribution order-disorder transformation results in Fe$_{75}$Ge$_{25}$ alloys, are presented. The disordered (A2) Fe$_{75}$Ge$_{25}$ [1] can be ordered in the DO$_3$ structure by thermal treating at 400 °C for 4 hs [2]. The Mössbauer study of the system treated at different temperatures between 100 and 400 °C allowed the investigation of the local atomic configuration around the Fe probes. It was found that the hyperfine field and isomer shift values depend essentially on the number of nearest neighbours. The short range order parameter, obtained by analysis of the Mössbauer spectra, undergoes an important increment at 200 °C. This analysis indicated that the transformation occurs by gradual ordering into the DO$_3$ structure. The XRD results, instead, show that long range order into the DO$_3$ structure begins at about 325 °C.

Introduction

In diluted Fe$_{100-x}$M$_x$ solid solutions the variation of the hyperfine field and isomer shift at [57]Fe probes can be represented as a linear function of the number of first and second neighbours M. Several investigations have employed this relation in iron base alloys up to 8 at. % of M[3]. This relation has also been used successfully for concentrations up to 18 at. % of germanium [4]. However in the Fe-Si system [5] it was found that the variation of isomer shift with the number of near neighbour is not linear.

In the present work, we present a detailed study of the variation of the Mössbauer parameters with the number of the near and next near neighbours germanium to the Fe probe, during the thermally activated transformation from the A2 disordered state to the ordered DO$_3$ one in Fe$_{75}$Ge$_{25}$. For this purpose a model which relates the probabilities of having n first neighbours M and m second ones, P(n,m), to the order parameter s of the DO$_3$ structure is used.

Experimental

Samples were prepared by mechanical alloying high purity elemental powders of Fe and Ge with composition 25 at. % of Ge. By alloying a solid solution bcc-Fe$_{75}$Ge$_{25}$ (A2) is obtained. The alloys were then thermally treated in vacuum for 4 hs between 100°C -400°C. The strain relaxation, crystallite growth, and chemical order development were measured by X-ray diffractometry and Mössbauer spectrometry. The X-ray characterisation was performed at the LNLS (Laboratorio Nacional de Luz Sincrotrón), Campinas, Brasil with $\lambda = 1.7652$ Å.

Results

In a previous communication a preliminary analysis of Mössbauer spectra by means of a hyperfine field (B) distribution, where a linear correlation between B and the isomer shift (δ) was allowed, was reported [2]. In this contribution we present an analysis of the Mössbauer spectra based in the atomic configurations around the Fe probes. Usually for diluted Fe$_{100-x}$M$_x$ solid solutions a correlation between B and δ at the Fe site on one side, and its solute near (n) and next near (m) neighbours on the other side, is supposed resulting in the following equation:

$$B(n,m)=B_0+B_1 n+B_2 m$$

$$\delta(n,m)=\delta_0+\delta_1 n+\delta_2 m$$

B_0 and δ_0 are the hyperfine field and the isomer shift corresponding to n=m=0 and B_i and δ_i with i=1,2 the contributions corresponding to one first or one second neighbour. These expressions were used in solid solutions of up to about 18 at. % of Ge [3]. In previous investigations [1,2], it was observed that the hyperfine field and isomer shift variations were not linear with the number of Ge first and second neighbours (see Fig.1).

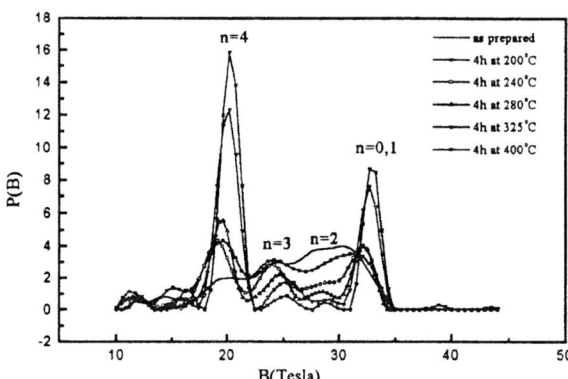

Fig.1: Hyperfine field distribution obtained form the Mössbauer spectra [1,2]

From the analysis of alloys fully ordered in the DO_3 structure with 18, 20 and 25 at. % of Ge, it was concluded that the variation of B and δ with n and m follows the trends as shown in figure 2:

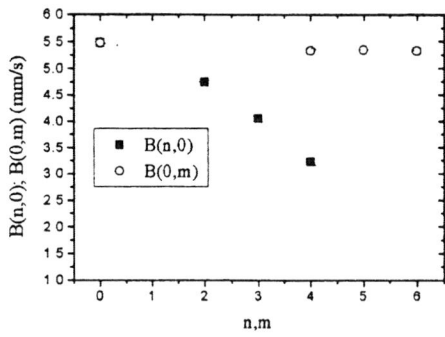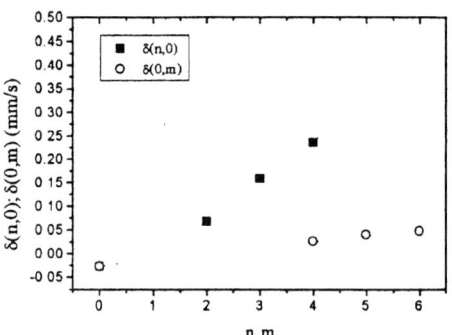

Fig.2: Evolution of the hyperfine field and isomer shift with the number of first and second neighbours

$$B(n,m)=B_0+B_1 n^{3/2}+B_2 m^{1/2}$$
$$\delta(n,m)=\delta_0+\delta_1 n^{3/2}+\delta_2 m$$

Figure 3 shows the Mössbauer spectra with the T_{ann} = 100-400 °C A2-> DO_3 evolution induced by thermal treatments in the temperature range. It is possible to establish a correlation between the P(n,m) probabilities and the DO_3 short range order parameter s. A fitting routine based on the variation of this parameter was constructed and the results obtained with this routine are shown in Fig. 3, while the short range order evolution with the temperature is presented in Fig. 4. The initial state of the system shows a partial degree of order (s=0.4) and remains until treatment temperatures of about 150°C are reached. An important change of s was observed between 150 and

A.F. Cabrera et al. / Evolution of the short range order parameters for the transformation A2 $\Rightarrow$ D0$_3$ in Fe$_{75}$Ge$_{25}$

139

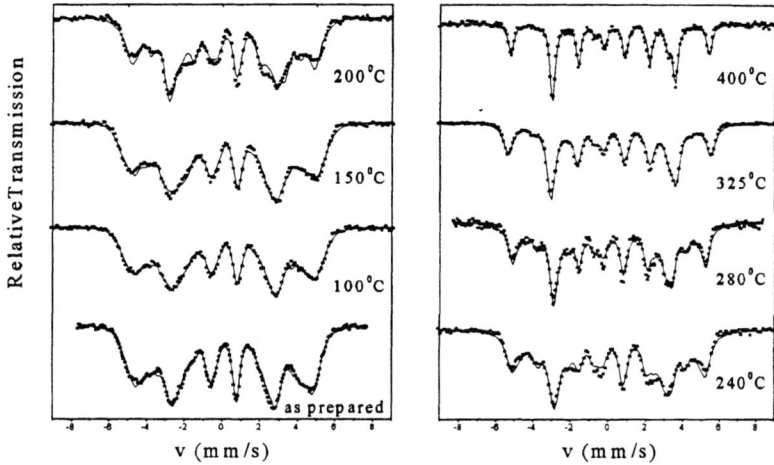

Fig.3: Mössbauer spectra of the mechanical milled Fe$_{75}$Ge$_{25}$ sample, as prepared and after annealing at various temperatures.

200°C reaching unity for T$_{ann}$ = 400 °C. The original values of the hyperfine parameters Bi, δi, were obtained from Figure 2. Although these parameters were set free during the fitting procedure, they remain approximately constant for all the samples with the exception of B$_0$ which increased from about 5.6 to 6.0 mm/s with s.

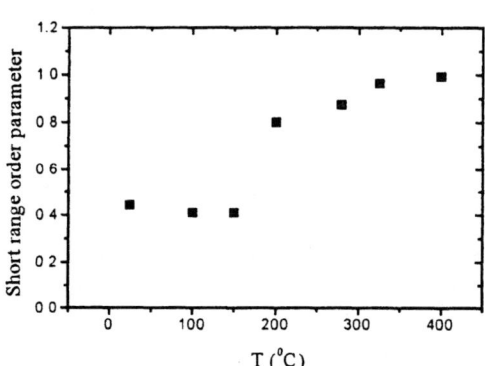

Fig.4: Evolution of s with the annealing temperature

The XRD patterns (see figure 5) show four peaks corresponding to the solid solution. These peaks became narrower and taller with increasing annealing temperature. For the 325°C case, more reflections appear, corresponding to the structure D0$_3$. A Rietveld refinement method, based on the D0$_3$ structure show that the crystallite growth and the strain relaxation evolve with increasing temperature. The grain evolution presents two steps. For temperatures below 325 °C, the most important feature is the increase of the short range order in the system, while between 325 and 400 °C an important increase of the D0$_3$ domain and/or crystallite size was observed. The strain decrease with increasing temperature indicates a relaxation within the particle. The lattice parameter (structure D0$_3$) obtained by Rietveld refinements presents a contraction with temperature for T$_{ann}$< 325 °C and remain approximately constant for T$_{ann}$> 325 °C. This may be another evidence of domain/crystalline growth during the last step. The volume contraction may

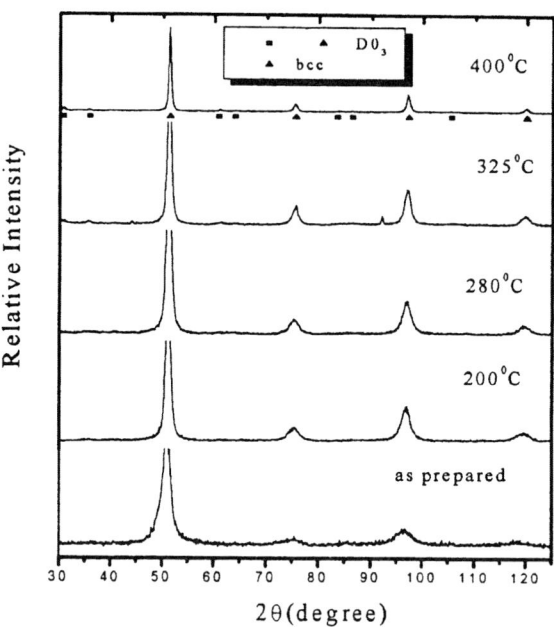

Fig.5: X-ray diffraction pattern of the as prepared and
annealed samples

be associated to the increment of the cohesion energy with increasing order, or to the elimination of point and extended defects as a consequence of the thermal treatments.

Conclusions

By thermal treating the metastable solid solution A2 (cI2-W) at temperatures between 100 and 400 °C the gradual ordering of the system into the metastable DO$_3$ structure (cF16-Li$_3$Bi) was observed. It was found that the hyperfine quantities δ and B can be reasonably described using a power law dependence on first and second neighbours to the Fe probes. A strong dependence on the number of near neighbours to the power 3/2 was encountered in both cases, indicating the dependence on the number of second neighbours much weaker and not relevant. The short range order parameter s varied between 0.4 (as prepared alloys and after annealing up to 150 °C) and 1.0 (after treatment at 400 °C). From above 325 °C the occurrence of long range order was detected. This ordering was coincident with the observation of an important increase in domain/crystal size. The lattice parameter and mean square strain decreased monotonically with treatment temperature.

References
[1] A.F. Cabrera, F.H. Sánchez and L. Mendoza Zélis, Hyp. Int. C 2 (1997) 230
[2] A.F. Cabrera, F.H. Sánchez and L. Mendoza Zélis, International Conference on the Applications of the Mössbauer Effect. Brasil, September 1997 (to be published)
[3] I. Vincze and T. Aldred, Phys. Rev. B 9 (9) 1974 3845
[4] L. Brossard, G.A. Fatseas, J.L. Dorman and P. Lecocq, J. App. Phys. 42 (4) (1971) 1306
[5] M. B. Stearns, Phys. Rev 129(3) (1963) 1136

MÖSSBAUER STUDIES AND NORMAL STATE MAGNETISM OF CaLaBaCu$_{3-x}$Fe$_x$O$_{7-\delta}$ SUPERCONDUCTOR, 0.00≤x≤0.21

Angel Bustamante Dominguez[1], D. A. Landinez Tellez [2] and J. Albino Aguiar[2].

[1] Facultad de Ciencias Físicas, Universidad Nacional Mayor de San Marcos,
Apartado Postal 14-0149, Lima-14, Perú.
[2] Departamento de Fisica, Universidade Federal de Pernambuco, 50670-901 Recife PE, Brazil.

The synthesis of the compounds CaLaBaCu$_3$O$_7$ (CLBO) has provided a new high-temperature superconducting system with unusual properties. The material is a tetragonal bulk superconductor with a T$_c$ of 80K, and show no evidence for a/b twining. CaLaBaCu$_3$O$_7$ superconductor is a prototype of the Y$_1$Ba$_2$Cu$_3$O$_7$ (Y:123) structure, in which the Y^{3+} and Ba^{2+} sites have been partially replaced by Ca^{2+} and La^{+3} ions. We report here results obtained by ac susceptibility and dc magnetization measurements and the Mössbauer spectroscopy for samples CaLaBaCu$_{3-x}$Fe$_x$O$_7$ with 0.00≤x≤0.21. The critical temperature T$_c$ for these systems is defined as the onset of the diamagnetic transition. For the pristine sample (x=0.00) this value is 79K and decreases with increasing Fe substitution at the Cu sites until T$_c$=39 K for x=0.21. The Mössbauer spectra of the samples at room temperature (RT) showed four species labeled **A, B, C, E** and **F**. Each of these species correspond to different oxygen configurations for Fe in the Cu sites similarly to iron doped with (Y:123) compounds. The normal state high field (3 kOe) dc magnetism of these samples infers a nearly temperature independent and relatively small molar susceptibility for the x=0.00 sample, while for the Fe doped samples the same follows a Curie-Weiss temperature dependence, in terms of a localized moment presumably on doped Fe ions.

1. INTRODUCTION

After the discovery superconductivity exists as a bulk phenomenon at 91K in Y$_1$Ba$_2$Cu$_3$O$_7$ has stimulated extensive studies on the effect of substitutions on the superconducting properties of this compound [1, 2]. CaLaBaCu$_3$O$_7$ which is one of (Y:123) like compounds is a bulk superconductor with a T$_c$ of 80 K [3-5]. CLBO is isomorphic to tetragonal (Y:123), in which the Y^{3+}-site is occupied by the Ca^{2+} and La^{+3} ions and the Ba^{2+}-site is occupied by the Ba^{2+}, La^{+3} and Ca^{2+} ions [4,5]. It is observed by transmission-electron-microscopy micrographs on CLBO that show no evidence for a/b twining, which is observed in Y$_1$Ba$_2$(Cu$_{1-x}$Fe$_x$)$_3$O$_7$ [6]. Thus CLBO with T$_c$=80 K has a tetragonal crystal structure with a=3.8694Å and c=11.6138Å and space group P4mm and has no oxygen deficiency (δ∼7) [4,5].

We found that there are only few reports on the 3d metal substitutions in the Ca based CLBO compound [7,8,9,10]. In this work we present the results of ^{57}Fe substitution at the Cu sites in the CaLaBaCu$_3$O$_7$ compound; its site occupancy at Cu(1) in the chains and Cu(2) in the planes are studied by Mössbauer spectroscopy at RT. The normal state magnetism of the Fe doped CaLaBaCu$_3$O$_7$ compounds is studied by magnetization measurements both at superconducting and normal states. The normal state dc susceptibility follows a Curie-Weiss behavior in terms of the localized moment of Fe ions, similar to that reported previously for the Fe doped orthorhombic RE : 123 systems [11]

2. EXPERIMENTAL

Samples with the nominal composition CaLaBaCu$_{3-x}$Fe$_x$O$_7$ (where 0.00≤x≤0.21) were prepared by thoroughly mixing and grinding of La$_2$O$_3$, BaCO$_3$, CaCO$_3$, CuO and ^{57}Fe$_2$O$_3$ in stoichiometric ratio and firing at 950°C for 12 h. The calcinated material was regrounded and calcination process was repeated three times. After the calcination the material was reground, pelletized and annealed at 975°C for 24 h in oxygen atmosphere. The samples were cooled down to 575°C slowly and kept at this temperature for 24 h, at the same ambient conditions. Finally, the samples were slowly cooled down to RT.

The X-ray diffraction (XRD) analysis were performed in an automatized Siemens Type-F diffractometer in a scanning mode θ-2θ with steps of 0.025°. A graphite monocromator was used to select the Cu-Kα doublet of the diffracted radiation.

AC susceptibility measurements were done on a Quantum Design SQUID magnetometer with an ac field amplitude of 0.1 Oe and frequency at 31 Hz, for temperature between 100 and 5 K. For the normal state dc magnetism, the samples were also studied in a dc field of 3 kOe in the temperature range of 80 to 300K

The Mössbauer absorption spectra on ^{57}Fe were taken at RT in transmission geometry, using a ^{57}Co/Rh source with a weight absorber of 100 mg. The spectrometer was operated with a triangular source motion and the spectra were accumulated with a 512 multichannel analyser. Calibration was performed with a foil metallic.

3. RESULTS AND DISCUSSION

The powder XRD pattern for all samples confirm that the diffraction lines can be indexed with a tetragonal unit cell with space group P4/mmm that is similar to the prototype pure (Y:123) like compounds. The ac susceptibility results for all the samples of CaLaBaCu$_{3-x}$Fe$_x$O$_7$ with 0.00≤x≤0.21 are shown in Fig. 1 and confirming the superconductor behavior. The critical temperature T$_c$ for these systems is defined as the onset of the diamagnetic transition. For the pristine sample (x=0.00) this value is 78 K. The transition temperature depends in general on the ratio La/Ca in the general formula La$_{3-x}$Ca$_{2-x}$Ba$_{3-x}$Cu$_6$O$_y$ [12-14]. For the present composition with La/Ca=1, i.e. x=1, the reported T$_c$ values range from 68 to 80 K [5, 12-14, 16,]. With increasing Fe substitution at the Cu site T$_c$ decreases linearly. The T$_c$ values are showed in Table 2. It has been shown earlier for Fe doped RE:123 systems that T$_c$ of the pristine system decreases with iron doping [17].

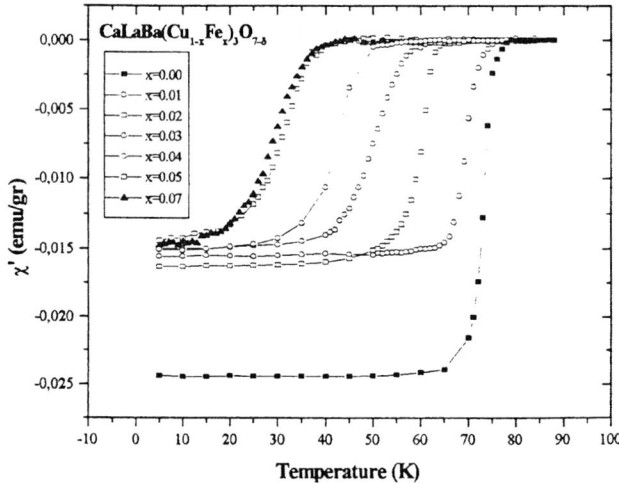

Fig. 1. Ac susceptibility versus temperature plots for CaLaBaCu$_{3-x}$Fe$_x$O$_7$ compounds with different x values.

The decrease in T$_c$ of the doped system can be explained in three possible ways. First, the Fe^{3+} substitution at Cu^{2+} sites decreases the number of mobile carriers, due to hole filling, whereby T$_c$ is reduced. Second, the decrease in T$_c$ can be due to the disorder created by the Fe ions in the unit cell of the pristine system. The extent of disorder in a clean system increases with ionic size mismatch between the host ion and the impurity. The ionic size mismatch of host Cu^{2+} with the impurity Fe^{3+} ion can be the another important cause for the decrease in T$_c$ with the above substitution. Finally, the magnetic moment of Fe^{3+} can be the third reason for T$_c$ depression : We will consider this aspect later after discussing the results of Fig. 4.

A typical Mössbauer spectra of CaLaBaCu$_{3-x}$Fe$_x$O$_7$ at RT is given in Fig. 2 with x=0.03, 0.06 and 0.09 and in the Fig. 3 for x=0.12, 0.15 and 0.21 which represents 1, 2, 3, 4, 5 and 7 at. % of iron occupying the Cu sites. We used a fitting model restrained by a single value of the linewidths for the four symmetrical doublets, labelled **A**, **B**, **C**, **E** and **F** corresponding to Fe in different local environments and did not show any magnetic component at high velocity in contrast with the reported by E. R. Bauminger et. al. [18]. The corresponding Mössbauer hyperfine parameters are listed in Table 1.

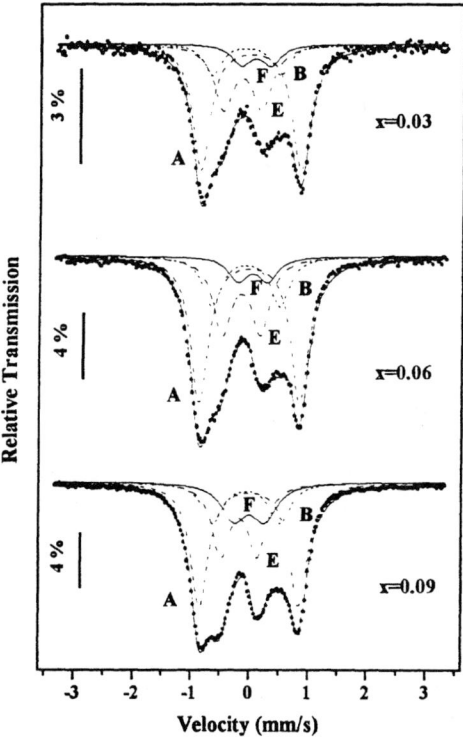

Fig. 2. Mössbauer spectra at RT of sample CaLaBaCu$_{3-x}$Fe$_x$O$_7$ with x=0.03, 0.06 and 0.09.

The isomer shift (δ) and quadrupole splitting (Δ) values of doublets **A**, **B**, **C**, **E** and **F** are similar to those reported in the literature for the Y:123 system [17, 18, 19]. The species **A** predominates in all spectra. It corresponds to Fe^{3+} ions in Cu(1)-chains positions with a fourfold (plane square) oxygen coordination. The species **B** is assigned to Fe^{3+} substitution at the Cu(1) site in fivefold pyramidal oxygen coordination. The isomer shifts of species **A** and **B** indicate the iron state is trivalent in the intermediate-spin state (S$_z$=3/2). The species **C** corresponding to substitution of Fe in the Cu(2) site in quasi-octahedral oxygen coordination, the isomer shifts corresponding to high-spin state (S_z=5/2) Fe^{3+} ion. This fact is due to the presence of Ca^{2+} near the Cu(2) planes. The species **E** is typical of Fe clusters in Cu(1) sites [20]. The species **F** corresponding to substitution of Fe in the Cu(2) site in fivefold pyramidal oxygen coordination, the isomer shifts corresponding to high-spin state (S_z=5/2) Fe^{3+} ion.

Table 1
Mössbauer parameters at RT for the different species in the CaLaBaCu$_{3-x}$Fe$_x$O$_7$ system doped with iron. δ=Isomer Shift relative to metallic Fe at RT ($\pm$0.002mm/s), Δ=Quadrupole Splitting ($\pm$0.01mm/s), Γ=Full width at half maximum ($\pm$0.01mm/s) and A=Relative area ($\pm$3%).

Samples	A				B				F/C				E			
	δ	Δ	Γ	A	δ	Δ	Γ	A	δ	Δ	Γ	A	δ	Δ	Γ	A
x=0.03	0.101	1.70	0.36	51	0.021	1.17	0.35	12	0.182	0.51	0.40	09	-0.044	0.65	0.40	28
x=0.06	0.104	1.71	0.37	48	0.037	1.16	0.38	17	0.155	0.51	0.40	09	-0.037	0.65	0.38	26
x=0.09	0.115	1.69	0.37	43	0.068	1.15	0.42	16	0.120	0.53	0.50	17	-0.055	0.62	0.36	24
x=0.12	0.104	1.72	0.38	46	0.060	1.16	0.37	15	0.155	0.55	0.40	09	-0.032	0.64	0.44	30
x=0.15	0.125	1.70	0.36	38	0.114	1.18	0.43	16	0.341	0.45	0.60	08	-0.019	0.58	0.45	38
x=0.21	0.122	1.70	0.36	27	0.142	1.20	0.50	23	0.278	0.45	0.60	23	-0.033	0.57	0.45	27

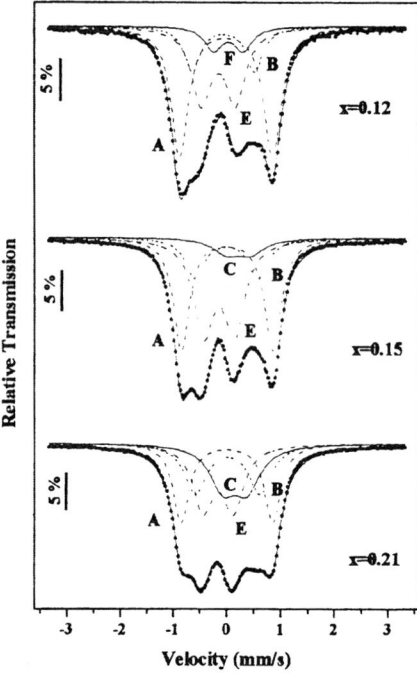

Fig. 3. Mössbauer spectra at RT of sample CaLaBaCu$_{3-x}$Fe$_x$O$_7$ with x=0.12, 0.15 and 0.21.

The most interesting observation in Table 1, is the increase of the population of the species **C** (~37%), when iron is increase, at expense of doublets **A** and **B**, originating a migration of iron from the Cu(1) chains into the Cu(2)O$_2$ planes and lead to a decrease of both the transition temperature and superconducting volume. The population of species **E** practically has not changed with increasing Fe content. The line widths of species **F** and **C** exceed that of the remaining ones by a factor of 1.5. This can be explained by the following causes : (i) by the disordering in the vacancy distribution around the resonance atoms, which leads to a EFG-distribution, (ii) by the summation of the unresolved components corresponding to the 5-multiplicity oxygen atom coordination of the square pyramid, which can exist in the Cu(1) and Cu(2) positions [21].

All the samples in their normal state, i.e. above 80K, follow the well known Curie-Weiss relation $\chi = \chi_0 + C/T - \theta_p$ where χ_0 is the temperature independent susceptibility, C is the Curie constant and θ_p the Curie temperature. By fitting the data with the above expression one can get the effective paramagnetic moment μ_{eff} of the impurity ion (Fe) in the system. It is worth mentioning that none of the ions of the unit cell of (CLBO) give rise to any temperature dependence of the susceptibility, as discussed in earlier report [9,10]. Further, our samples do not contain any impurity phase. Hence we conclude that the observed temperature dependence of the magnetic susceptibility in Fig. 4 is due to the substituted Fe ions only. The fitting parameters were calculated from the molar susceptibility of the samples and then by assuming that the whole moment is due to Fe only, the contribution was calculated per ion..

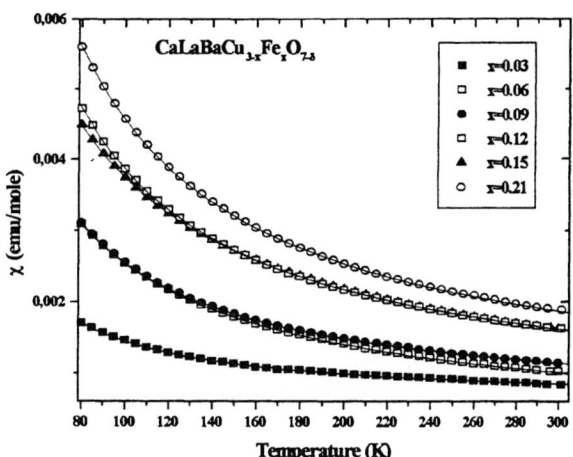

Fig. 4 Plots of normal state dc magnetic susceptibility in a field of 3 kOe versus temperature for CaLaBaCu$_{3-x}$Fe$_x$O$_7$ compounds

Table 2. Transition temperature (T$_c$), magnetic properties (θ_p, χ_0, C) and Fe moment for CaLaBaCu$_3$-$_x$Fe$_x$O$_7$ system.

Samples	T$_c$	χ_0 (emu/mol)*10^{-3}	C (emu*K/mol)	θ (K)	μ_{eff}	Range of Temp.
Fe1%	74	5.9	0.07466	13.4	4.460	80-300
Fe2%	64	2.2	0.2419	-4.45	5.676	80-300
Fe3%	59	4.9	0.19255	6.52	4.1344	80-300
Fe4%	50	4.6	0.35831	-9.11	4.885	80-300
Fe5%	40	5.5	0.31713	-4.25	4.110	80-300
Fe7%	39	5.7	0.3860	-7.44	3.832	80-300

The fitting parameters obtained from the above exercise are tabulated along with T$_c$ in Table 2. The μ_{eff} decrease systematically with the iron concentration, a similar type of trend was also observed earlier in Fe doped RE :123 systems [11]. The possible explanation for this is the crystal field effect which screens the effective moment of the magnetic ion. This effect increases with an increase in the concentration of magnetic impurity.

Acknowledgement

The authors thank Mössbauer Spectroscopy Group of Brazilian Center for Research in Physics (CBPF) for assistant the ^{57}Fe$_2$O$_3$ compounds and Manoel Rothier at the Universidade Federal de Rio de Janeiro, Brasil, for the X-rays diffraction. The present work was financed by the Brazilian Agencies CNPq, FACEPE and FINEP. D. A. Landinez Tellez thanks CNPq and Colombian Science Agency.

REFERENCES

[1] A.V. Narlikar, S.K. Agarwal and C.V. Narsimha Rao, in : Studies of high temperature superconductors, de. by A. V. Narlikar, (Nova Science Publishers, New York, 1989) p. 343.

[2] J.M. Tarascon and B.G. Bagley, in : Chemistry of high temperature superconductors, ed. A. Vanderaha (Noyes Publishers, New York, 1993) p. 310.

[3] Y. Tokura, J. B. Torrance, T. C. Huang and A. I. Nazal, Phys. Rev. B 38 (1988) 7156.

[4] D. M. de Leeuw, C. H. A. Mutsaers, H. A. M. van Hal, H. Verweij, A. H. Carim and H. C. A. Smoorenburg, Physica C 156 (1988) 126.

[5] W. T. Fu, H. W. Zandbergen, C. J. van der Beek and L. J. de Jongh, Physica C 156 (1988) 133.

[6] Y. Syono, M. Kikuchi, K. Oh-ishi, K. Hiraga, H. Arai, Y. Matsui, N. Kobayashi, T. Sasaoka and Y. Muto, Jpn. J. Appl. Phys. 26 (1987) L498.

[7] T. Watanabe, K. Kawase, I. Shiono, H. Yumoto, J. Furukawa and T. Tsukamoto, Physica C 185-189 (1991) 557.

[8] R. Singh, R. Lal, U. C. Upreti, D. K. Suri, A. V. Narlikar, V. P. S. Awana, J. Albino Aguiar and Md. Shahabuddin, Phys. Rev. B 55 (1997) 1216.

[9] V. P. S. Awana, D. A. Landinez Tellez, J. M. Ferreira, J. Albino Aguiar, R. Singh and A. V. Narlikar, Mod. Phys. Lett. B 10 (1996) 619.

[10] V. P. S. Awana, Rajvir Singh, D. A. Landinez Tellez, J. M. Ferreira, J. Albino Aguiar and A. V. Narlikar, Physica C 277 (1997) 265.

[11] V.P.S. Awana, R. Lai, D. Varandani, A. V. Narlikar and S.K. Malik, Supercond. Sci. Tech. 8 (1995) 745.

[12] R. A. Gunasekaran, I. K. Gopalakrishnan, P. V. P. S. S. Sastry, J. V. Yakhmi and R. M. Iyer, Physica C 199 (1992) 240.

[13] S. Engelsberg, Physica C 176 (1991) 451.

[14] Y. D. Leu, W. N. Huang, C. M. Wang and H. C. I. Kao, Physica C 261 (1996) 284.

[15] D. S. Wu, Y. F. Yang, H. C. I. Kao and C. M. Wang, Physica C 212 (1993) 32.

[16] J. L. Peng, P. Kalvins, R. N. Shelton, H. B. Radousky, P. A. Hahn, L. Bermudez and M. Costantino, Phys. Rev. B 39 (1989) 9074.

[17] E. Baggio-Saitovitch, I. Souza Azevedo, and R. B. Scorzelli, H. Saitovitch, S. F. da Cunha, A. P. Guimarães, P. R. Silva, and A. Takeuchi, Physical Review B 37 (1988) 7967.

[18] E. R. Bauminguer, D. Edery, I. Felner and I. Nowik, Hyperfine Interactions 55 (1990) 1195.

[19] P. Boolchand, and D. McDaniel, in: Studies of High Temperature Superconductors, vol 4 ed. A. V. Narliker (Nova, New York, 1990) p. 143.

[20] H. Oesterreicher, M. Smith and D. Taylor, J, Mag. Mag. Mat. 104-107 (1992) 497.

[21] V.A. Virchenko, V.S. Kuz'min, T.M. Tkachenko and A.V. Shablovskii, J. of Mag. and Mag. Mat. 183 (1998) 78-80.

NEW INFORMATION SYSTEMS FOR MÖSSBAUER SPECTROSCOPY: A STATUS REPORT FROM THE MÖSSBAUER EFFECT DATA CENTER

J. G. Stevens, A. M. Khasanov, E. J. Kolb and R. L. Burgess

Mössbauer Effect Data Center, University of North Carolina, Asheville, NC 2880-85114
United States

The Mössbauer Effect Data Center has been providing information services to the international Mössbauer community for almost thirty years. There are currently almost 40,000 bibliographical references from which the Center has abstracted over 75,000 data entries. Latin American scientists have become major contributors in the field of Mössbauer spectroscopy as documented by the Center. With recent developments in database management and the World Wide Web, the Center has undergone dramatic changes in its methods of providing information. An experimental program exclusively for Mössbauer researchers residing in Latin American countries has been developed. This program is expected to greatly increase the amount of Mössbauer spectroscopic information available to Latin American researchers, while dramatically reducing the cost for providing such services.

1. Mössbauer Community

Forty years ago Professor Rudolf L. Mössbauer published a paper in *Z. Phys.* [1] which initialized a new research area that has impacted many segments of the sciences. Research in this area became known as the Mössbauer effect, and more recently as Mössbauer spectroscopy, and has encompassed an international scientific community unto itself. Now, forty years since the original publication, there are almost 40,000 publications on the Mössbauer effect. The growth in the number of publications is given in Figure 1 in which the number of publications per year is plotted up through 1997. Several items of particular note are that in recent years there have been between 1,500 and 1,600 publications per year of which approximately 500 per year of these on the average are coming from special conference publications.

The Mössbauer Effect Data Center has in its database approximately 30,000 authors coming from 107 countries with publications in 46 different languages. These Mössbauer papers have been published in over 2,700 different journals and 1,000 different books. While many, if not most of the very early publications were focused on the understanding of the Mössbauer effect, today almost all publications are focused on using the Mössbauer effect (Mössbauer spectroscopy) to investigate various materials and systems.

2. Latin American Mössbauer Community

Geographically the fastest growing segment of the Mössbauer community today is the Latin American Mössbauer community. For the last ten years the Latin American Mössbauer community has been contributing 60 to 120 publications per year with a large fluctuation from one year to another as a result of the publication of the proceedings of the LACAME conferences every two years. Brazil continues to remain the dominant contributor from Latin America with approximately two-thirds of the Mössbauer publications. The percentage of contributions from the Latin American countries is shown in Figure 2. Of particular note is the maturity that has taken place in the Latin American Mössbauer community which is borne out by the increase of total publications over the last ten years which has tripled. Of additional note is a comparison of a report at the 1992 LACAME, in which there were three individuals in Latin America having 50 or more publications [2]. Today that number has increased to seven. There are 23 individuals having 25 or more publications. Their names are given in Figure 3. Countries with scientists represented in this list include Argentina, Brazil, Mexico and Venezuela.

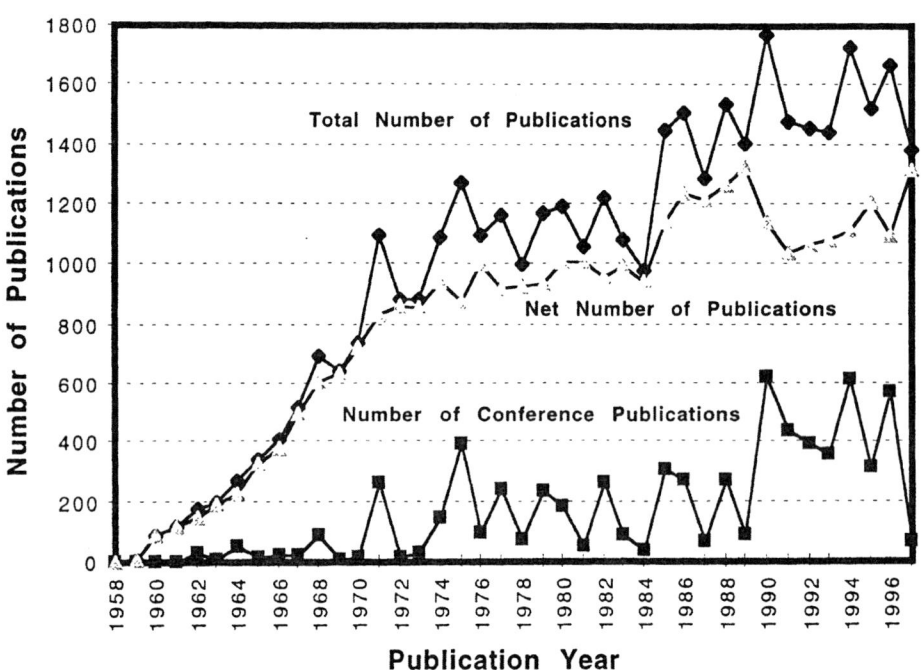

Fig. 1. The growth in Mössbauer spectroscopy since the original publication of Rudolf L. Mössbauer.

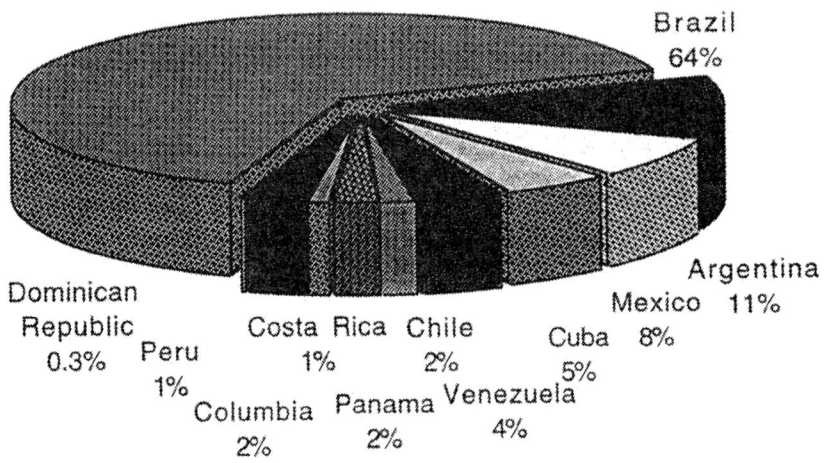

Fig. 2. Distribution by Latin American countries giving percentage of the total number of Mössbauer publication from each country.

Quarter-Century Latin American Club		
Scientist	**Country**	**Number of Publications**
E. M. Baggio-Saitovitch	Brazil	113
J. Danon	Brazil	104
V. K. Garg	Brazil	74
R. B. Scorzelli	Brazil	63
A. Vasquez	Brazil	56
H. R. Rechenberg	Brazil	56
R. C. Mercader	Argentina	50
H. D. Pfannes	Brazil	49
E. Galvao da Silva	Brazil	46
F. Gonzalez-Jimenez	Venezuela	39
I. J. R. Baumvol	Brazil	37
I. Souza Azevedo	Brazil	34
L. Amaral	Brazil	34
M. Boge	Brazil	34
D. Guenzburger	Brazil	33
C. Saragovi	Argentina	32
F. H. Sanchez	Argentina	32
A. Abras	Brazil	31
W. H. Schreiner	Brazil	29
S. K. Xia	Brazil	28
S. Aburto	Mexico	26
M. Jimenez	Mexico	25
C. A. Taft	Brazil	25

Fig. 3. Listing of Latin American scientists having 25 more publications according to the database files of the Mössbauer Effect Data Center.

3. Mössbauer Effect Data Center

The Mössbauer Effect Data Center has been serving the Mössbauer community for over 30 years and its history has paralleled very much the development of computer information systems. The Center has experimented over the years with a variety of information systems to provide for the needs of the Mössbauer community. It is interesting that the information resource that is still the most widely used is the *Mössbauer Effect Reference and Data Journal* which is published ten times a year. This journal, while having experienced decreases in its number of subscriptions (which parallels that of other scientific journals internationally) is still by far the most widely used of the various information services of the Center. The Center also provides a series of other services including printed handbooks, small subsets of the Center's database on disks referred to as *Micros*, and more recently the establishment of both a homepage and access to the Center's main database through the Internet which is described in more detail below.

The Center is comprised of four individuals, only one of whom is employed full-time by the Center. Additional assistance is provided to the Center by both a group of Associate Editors numbering 12, and 20 members of the International Advisory Board. The Associate Editors provide assistance in obtaining Mössbauer publications from around the world. They additionally assist in abstracting the data information from these publications for storage in our databases. Both the Associate Editors and International Advisory Board provide direction and support for the Center.

4. Two-Year Plan for Latin American Scientists

The Mössbauer Effect Data Center is proposing to initiate a two-year experiment to drastically reduce the subscription cost of the *Mössbauer Effect Reference and Data Journal* for scientists in Latin American countries. Scientific journals over the last 10 to 15 years have been increasing at the rate of 10-20% or more per year as a necessity created by a number of economic forces. The Mössbauer Effect Data Center is proposing an experiment to decrease by 50% the subscription cost versus the 10-20% increase trend. In particular for 1999, the regular cost for a subscription of the *Mössbauer Effect Reference and Data Journal* is US$700--the *Journal* will be offered to Latin American scientists and institutions for US$350. This is a two-year experiment because of the economic factors of this plan. Specifically, the number of subscriptions from Latin America will need to more than double during this period of time. This is being announced at LACAME '98 and is a challenge to the Latin American Mössbauer community. The most important aspect of the plan is an effort to provide the best information resources to a considerably larger group of scientists in Latin America. This, we envision, will further strengthen the development of Mössbauer spectroscopy in the Latin American countries and improve the quality of contributions Latin American scientists are making to the international science community.

5. Internet Access to MEDC Database

Beginning in January 1999, subscribers to the *Mössbauer Effect Reference and Data Journal* may gain access to the Center's database, containing both bibliographic and data information, for a small additional fee of US$100. The bibliographic part of the database contains almost 40,000 references, while the data portion contains over 75,000 sets of data. Both portions contain keywords which assist in identifying the reference and/or data needed. Included with the references are article titles, all authors, journal name, volume, page and year information. The data includes, in addition to the name of the material, isomer shifts, quadrupole splitting, temperature, and reference materials (see Figure 4).

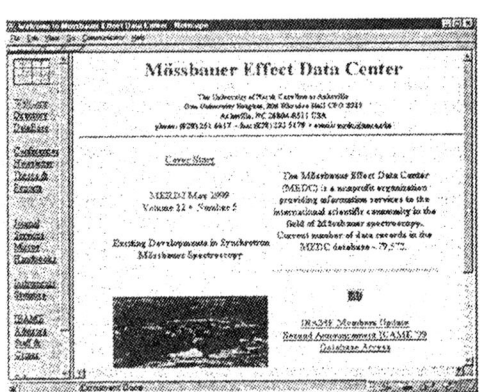

Fig. 4. Homepage of the Mössbauer Effect Data Center.

References

[1] R.L. Mössbauer, Z. Phys 151 (1958) 124.
[2] J.G. Stevens, Hyperfine Interact. 83 (1994) 125.